Production: H. B. Maynard and Company

Other titles in the McGraw-Hill European series in Management

Basil & Cook · The Management of Change
Ericsson · Materials Administration
Drake & Smith · Behavioural Science in Industry
Humble · Management by Objectives in Action
Humble · Improving the Performance of the Experienced Manager

PRODUCTION

an international appraisal of contemporary manufacturing systems and the changing role of the worker

H. B. Maynard and Company
Edited by Rolf Tiefenthal

A HALSTED PRESS BOOK

John Wiley and Sons
New York

Published in the U.S.A.
by **Halsted Press**
A Division of **John Wiley and Sons, Inc., New York**

Library of Congress Cataloging in Publication Data

Maynard (H. B.) and Company
Production: an international appraisal of contemporary manufacturing systems and the changing role of the worker

"A Halsted Press book."
1. Industrial management. 2. Industrial organization.
3. Personnel management. I. Tiefenthal, Rolf. II. Title.
HD31.M37 1975 658.5 75-16144
ISBN 0 470 58033 X

PRINTED AND BOUND IN GREAT BRITAIN BY
HAZELL WATSON & VINEY LTD, AYLESBURY, BUCKS

Contents

Preface vii

Acknowledgements viii

1. The Industrial Revolution writes a new chapter 1
2. The work and the worker 17
3. Designing production systems 28
4. Choosing the production mode 38
5. Planning the planning system 57
6. Structuring pay systems 78
7. Housing the system 103
8. Organizing for participation 112
9. Managing change (1) 132
10. Managing change (2) 140
11. Summary 153

Appendix. Some who tried 162

Index 177

Preface

In 1971, the International Academy of Management questioned 512 senior managers in 20 different countries, and asked them to identify the most crucial management problems that have not yet found satisfactory answers. Among those that were given the highest priority were:

> How can employees be motivated—i.e., how can we achieve a common understanding of those tasks and goals of mutual interest to all members of an organization, and how can a truly cooperative capability and will be established?

> How can the far-reaching decentralization that is seen to be necessary be achieved without risking loss of control and coherence?

This book is an attempt by H. B. Maynard and Co. to use the accumulated experience of its consultants over the last 40 years to demonstrate how these problems are manifesting themselves in today's industrial organizations. It indicates the danger of an unrealistic or merely defensive response, and suggests how an organization can ascertain and evaluate its options and implement its own solutions in a systematic fashion.

Acknowledgements

The conceptual planning, coordination and technical editing of this book was done by Rolf Tiefenthal, Principal, H. B. Maynard.

The material was transferred into a coherent and readable form by Graham Cleverley.

The Maynard consultants involved in the preparation of the material were: Benelux—Anton A. Smits and Jan Frederickse; Denmark—Willy Krøyer and Ove Nielsen; Finland—Professor Raimo Nurmi and Erkki Pursiainen; France—Claude Bonvalot and Christian P. de Malherbe; Germany—Ulrich B. Roedl; Italy—Cesare C. Mainardi and Pierluigi Matteucci; Sweden—Ingemar Johansson, Nils-Olaf Ljungqvist, Sture Selander, and Nils Vestergren; United Kingdom—George Fox.

Advisory committee: Laurens van den Muyzenberg (Chairman), Joseph J. Lands, Nils-Olaf Ljungqvist, and Erik Schou.

Others whom Maynard wishes to thank for their contributions, advice and assistance are: Rolf Lindholm (Swedish Employers' Confederation) who gave so freely of his time and supplied us with valuable material for chapter 6; Anders Berg and Bo Sundberg (Anders Berg Architectural Company, Stockholm), who contributed the brewery case study; Holger Wästlund and Sten-Göran Johansson (VBB Vattenbyggnadsbyrån, Stockholm) who contributed the material for chapter 7; James Conway and Bob Wright (Amalgamated Union of Engineering Workers); Andrew Robertson (Confederation of British Industry); Harold Stieglitz (The Conference Board, New York); Ray Wild (Administrative Staff College, Henley); A. W. Turner (Dowty Mining Equipment).

Janet Mead of Maynard acted as the administrative co-ordinator.

1. The Industrial Revolution writes a new chapter

Limits to the division of labour. Symptoms of discontent: the happy absentee. Education levels out. Who's afraid of the big, bad boss? The challenge to produce. Economic and human, too?

Two centuries ago, in 1776, Adam Smith published his 'Inquiry into the Nature and Causes of the Wealth of Nations', the most celebrated (if not necessarily the first) exposition of the principal doctrines of the phenomenon we have come to call the Industrial Revolution. By 1776, of course, that revolution had already been in progress for some while. Steam power had been harnessed to drive machines. Coal and iron were beginning to be seen as mineral resources potentially as valuable as gold and silver. The drift from a rural to an urban society had already begun. And, perhaps most importantly, new systems of industrial organization were already being developed.

But all those processes had begun spontaneously, pragmatically, not in response to a systematic doctrine. The doctrine was provided later, by Smith more than anyone.

Essentially, the doctrine had two apparent pillars (that it also masked some covert assumptions we will come to later). Enterprise should be free: and labour should be divided. Individuals and organizations should be left free to deal in open markets, taught Smith, because the invisible 'iron hand' of the market-place would ensure that in following their own self-interest, they would necessarily work to the public good.

And labour should be divided, because the process of division made tasks simpler, easier to learn and more speedily carried out: through division, the output of a given number of people in a given time could be immensely multiplied. Again, the idea was not new with Smith, nor even with the eighteenth century. But, as Schumpeter points out in his *History of Economic Analysis*, nobody before Smith 'ever thought of putting such a burden on the division of labour. With Adam Smith it is practically the only factor in economic progress. Alone it accounts "for the superior affluence and abundance commonly possessed even by the lowest and most despised member of Civilised society, compared with what the most respected and active savage can attain to".'

The division of labour in practice

At least in the liberal economies of the West—the ones we now call the 'developed' economies—Smith's doctrines were accepted, and, for a while at least, elevated into unchallengeable creeds. Helped by the growing development of technology, and the release of ever more mechanical energy to supplement muscle power (Smith would have argued that this was only made possible by the division of labour), affluence increased and spread. The charts in Fig. 1.1 demonstrate the results of the application of the doctrine in two sectors of the industrial world.

But though the Industrial Revolution had its triumphs, it also had its negative side. The 'Wealth of Nations' might be increasing, but not everyone was benefiting from it. As the nineteenth century wore on, contemporaries became more and more critical of the social effects of the system. In the 1830s, one Dr J. P. Kay wrote:

> But the population is crowded into one dense mass, in cottages separated by narrow, unpaved, and almost pestilential streets, in an atmosphere loaded with smoke and the exhalations of a large manufacturing city. The operatives are congregated in rooms and workshops during twelve hours in the day, in an enervating, heated atmosphere, which is frequently loaded with dust or filaments of cotton, or impure from constant respiration, or from other causes. They are engaged in an employment which absorbs their attention, and unremittingly employs their physical energies. They are drudges who watch the movements and assist the operations of a mighty, material force, which toils with an energy ever unconscious of fatigue. (Quoted by John Fielden in *The Curse of the Factory System*, 1836.)

Ceaseless pressure for social reform had its effect in Britain in, among other measures, the series of Factories Acts. Elsewhere it led to the emergence of Socialists like Saint-Simon, Marxism and revolutionary Communism, and paternalistic social welfare legislation of the sort associated with Bismark. That was on the practical side: on the theoretical side, even within the liberal tradition, by mid-century John Stuart Mill was driven to accept that the distribution of wealth, as opposed to its creation, was a matter of arbitrary political decision. Thus, within a century of the publication of the *Wealth of Nations*, the first chapter of the Industrial Revolution had ended: the doctrine of *laissez-faire* was, effectively, dead. Even in Britain, its lingering adherents weakened it into the doctrine of 'free trade'. It probably lasted longer than anywhere in the United States: but there it was assisted by the existence of the Frontier. The wide open spaces of the West, exploitable, not only in theory but also in practice, by any individual with the courage and initiative to do so, acted as a safety valve for frustration, and helped to foster the philosophy of individual self-reliance.

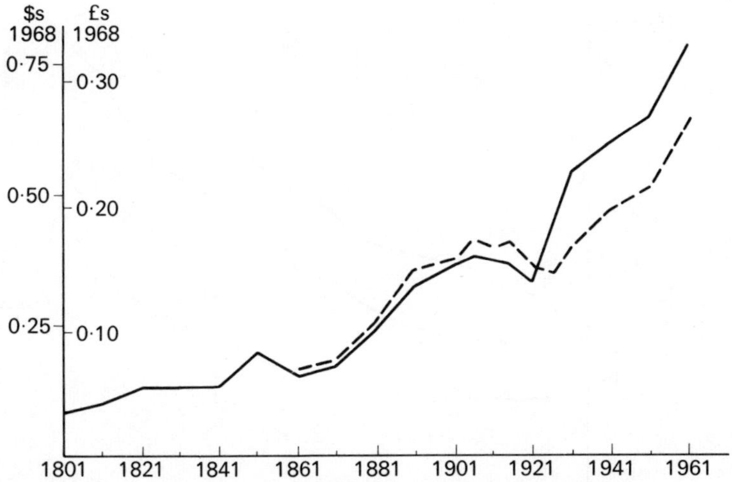

Fig. 1.1. 150 years of Adam Smith: the fruits of the Industrial Revolution. Purchasing power of one hour's basic wage, pre-tax, for skilled workers in the United Kingdom printing (solid line) and engineering (dotted line) industries. Cost of living indices only began to be compiled in the 1880s: before then calculations are based on the price of bread. Due to the massive increase in consumer choice, long-term comparisons are difficult. But decade on decade changes are reasonably comparable.

But if unlimited freedom and the pure profit motive were rejected, Smith's other pillar of liberal economics remained unchallenged and flourishing. Marxists, churchmen, romantics, humanists, might all speak out against unbridled capitalism, but nobody challenged the belief that the way to increased wealth for the whole community was through an ever more detailed application of the principle that jobs should be segmented as far as possible—that the way to increased productivity was to make tasks simple and speedily carried out.

Indeed, the principle gained new life in the early years of the twentieth century, when the methods and even the apparatus of the scientific laboratory began to be applied to the problem. Taking the methods of mechanical engineering as their model, F. W. Taylor, the Gilbreths and their followers used stopwatches and systematic analysis and measurement as a basis for job design in a way that justified the process being called 'industrial engineering'. Then came the assembly line.

And with the hot-house development provided by the two World Wars and the emphasis on economic growth that followed the Second, the principle reached its apogee.

Of course, the industrial engineers sometimes met with opposition. Many trades unionists might agree with the Swedish official who asserted: 'We

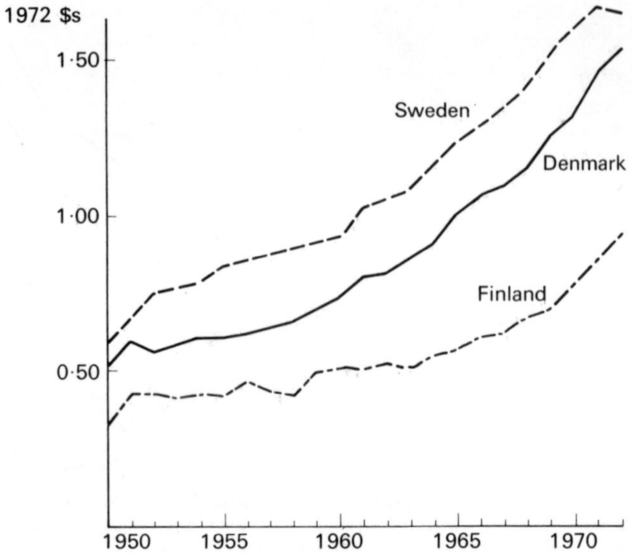

Fig. 1.2. The post-war pay-off: wages keep ahead of prices in Scandinavia. Purchasing power, one hour's basic wage, all industry, pre-tax.

want to get the biggest cake made first, *then* we'll argue about how it's shared'—but some were more concerned with *how* the cake was made. The American unions carried their opposition to Taylor himself as far as Congress. And in the 'forties and 'fifties, among trades unionists, the man with the stopwatch became a familiar bogey. The emissary of management with the singleminded task of squeezing more out of the workers, he was frequently, certainly in periods of industrial dispute, simply shut out of the plant.

But, by and large, this was part of the bargaining process. The point at issue was not whether industrial engineering in fact increased productivity: the point at issue was who should benefit from the increase. After all, the results were there to prove the merit of the techniques involved. In an inflationary era, prices rose continuously, but wages rose faster (Fig. 1.2). General living standards were rising at an unparalleled rate.

Moreover, rising productivity showed its benefits in other ways. Surplus income rose fast enough to allow governments in most countries to siphon some off in taxes (or 'insurance contributions') to provide increased security against old age, sickness and unemployment, and to increase investment in education. Again, productivity in the primary (raw material production, agriculture) and secondary (manufacturing) sectors rose enough to provide increasing incomes in the tertiary (services) sectors, and to allow the continual improvement of working conditions.

In their 'Agreement of Rationalization', signed on 1 June 1972, the Swedish Employers' Confederation and the Swedish Confederation of Trades Unions asserted: 'Considered in a broader perspective, rationalization constitutes the foundation for advances in all branches of economic activity as well as for economic progress generally. In this way rationalization helps to create the basis for successive improvements in the standard of living. At the same time rationalization is an important instrument in continuing efforts to improve the working environment and job satisfaction.'

In his report *On the Quality of Working Life* for the British Department of Employment, Dr N. A. B. Wilson could write:

> Despite intense concern over recent price increases, the increase in retail prices has consistently lagged behind the increase in weekly wage rates. Food prices have become less volatile while the range of choice has been greatly extended. Various items in the household budget previously classed as non-essential have now become essential and virtually universal. Paid holidays are now a requirement and are increasing in length. There is no doubt that similar kinds of progress could be demonstrated in other countries.
>
> At the workplace, also, there have been obvious advances. Medical, engineering and legislative efforts have transformed many, though not by any means all, industrial environments and practices from 'bad' to 'good' and often 'very good'.

Perhaps the picture was not as rosy everywhere. The late James Conway, general secretary of the Amalgamated Union of Engineering Workers, estimated in early 1974 that of the plants in which his members worked, one in four still had no toilets. Anywhere—in Germany, France, even the USA—one could find plants with tastefully decorated air-conditioned staff canteens, next door to workshops where the men ate their lunches off the workbench.

But, by and large, the Western countries each had in turn their 'economic miracles'. And the whole thing seemed to be firmly based on two things: ever more sophisticated technology, and ever more detailed and thorough-going application of Adam Smith's two-hundred-year-old principle of the division of labour.

Aesthetes and intellectuals, of course, had objections. They pointed out that, while physical exploitation had perhaps been eradicated, emotional and aesthetic exploitation remained. Camus wrote: 'Without work all life goes rotten. But when work is soulless, life stifles and dies.' When, at the end of *Modern Times*, Charles Chaplin and Paulette Goddard turn their backs on the assembly line and the audience, it is an indictment not of the profit motive, but of the mechanization of man. A few artists, like the Futurists, might revel in mechanization, but most, with Lawrence and Chesterton, bemoaned standardization and the dehumanization of industrial society.

But, like their predecessors from antiquity down to Rousseau and Byron, they mostly went unheard. A lot of people were ready to agree with the jibe: 'Alienation is what an intellectual feels when he sees someone doing a job of work.' Most people were not alienated by the system anywhere near as much as they were happy with the fruits of it. After the Second World War, in country after country, governments were re-elected on the slogan 'You never had it so good', demonstrating that politically material satisfaction still counted for more than aesthetic deprivation.

Then, however, things began to change.

Limits to the division of labour

It is never easy, of course, to pin down the beginnings of any historical change. Just as it is impossible to say that the Industrial Revolution started at such and such a time or such and such a place, so, too, is it impossible to state where and when, and in what way, it became obvious that a new chapter in the Industrial Revolution was beginning: a third chapter in which the principle of the division of labour would at last be challenged, not by romantics or artists, but by the very people whose concern was with increasing productivity.

The symptoms—strikes and absenteeism

The symptoms that something was changing, that something was wrong, are easier to trace. As an extreme example, strike action became more and more commonplace as the 'sixties came to an end. In Britain, there was a sudden and sharp take-off from a relatively quiescent level in 1968 that still shows no signs of stabilizing. The same year, of course, saw labour unrest bring down, eventually, the De Gaulle government. In 1967–71, the USA lost two-and-a-half times as many man-days in strikes as it had in 1960–66. And even havens of industrial peace like Sweden (especially in the winter of 1969–70) and West Germany (from 1971 on) saw their envied tranquillity seriously disturbed.

Less obtrusively, perhaps, reports of increasing absenteeism (other than for illness), labour turnover, and sabotage came from all over. When the Alcan Aluminium Corporation reported an industry average of 10% absenteeism, the response of many managers was that they could consider themselves lucky.

High turnover figures were also reported. In the UK telecommunications and electronics industries, Professor Ray Wild and Andrew Hill found turnover among female employees averaging 45% per annum, and reaching peaks as high as 65%, though, admittedly, this dropped away again as unemployment rose in the 'seventies. Almost comically, Texas Instruments in the USA recorded their turnover among cleaning staff at 100% a *quarter*.

In the passage already quoted from *On the Quality of Working Life*, Dr Wilson went on to sum up the situation:

> But in spite of these changes, or perhaps because they increase expectations, attitudes to work remain as problematic as ever. People show their dissatisfaction in complaints and formal grievances, in industrial disputes, in absences and unpunctuality, in a fairly widespread lack of commitment to work, and, in a small minority of cases, in calculation that no work or intermittent work would be preferable to continuous employment of the kind offered. Managers . . . are unhappy about a workforce much of which seems to lack not merely skill but a basis of reliability and work discipline; they are worried about a lack of concern for quality, and sometimes—so it seems to them—even of plain economic realism.

Very few people would disagree with his description, though there would be less unanimity about causes or solutions: suggested causes and proffered solutions have been the mainstay of most management literature in the past few years. But high among the propositions has been the one (which Herzberg was perhaps the first to put at the centre of his teaching) that, no matter how theoretically efficient, the division of labour is in the end self-defeating.

Simplification of tasks to lessen learning problems: fragmentation of work to enable it to be carried out more speedily: close specification of methods to ensure that operations are carried out in the most efficient way—the end result of all these may be not to segment the job, but to *fragment* it—to produce jobs that are inherently so meaningless and alienating that they destroy any commitment of the worker to the effectiveness of his work.

As General Motors found out after the massively publicized 'opening' of their plant at Lordstown, Ohio, in 1972, there is little point in having an ultra-modern, super-efficient factory in which no-one is working. In the UK, the International Publishing Corporation learned the same lesson from their equally sophisticated printing factory at Southwark. But, less extremely, there is also little point in establishing marginal increases in theoretical productivity if they are dwarfed by decreases in effort. As the authors of *Work in America*, the report specially commissioned by the US Secretary of Health, Education and Welfare, put it:

> What does the employer gain by having a 'perfectly efficient' assembly line if his workers are out on strike because of the oppressive and dehumanized experience of working on the 'perfect' line? As the costs of absenteeism, wildcat strikes, turnover and industrial sabotage become an increasingly significant part of doing business, it is becoming clear that the

current concept of industrial efficiency conveniently but mistakenly ignores the social half of the equation.

Increasingly sophisticated production design—even coupled with 'productivity bargaining'—no longer seems to bring commensurate results (Fig. 1.4). In the long run, Adam Smith appears to have been wrong, not only about free enterprise, but also about the division of labour.

At least in our society.

Underlying assumptions and changing attitudes

For, as we noted in passing earlier, there were some hidden assumptions in the philosophy of the Industrial Revolution: assumptions that no-one bothered to spell out, because everyone instinctively accepted them. And because, very likely, they were true at the time.

In essence, these assumptions would seem to have been:

(a) People will do as they are told by those in authority.
(b) People will accept that their superiors know better than they do.
(c) Man's prime motivation is economic gain.
(d) Man's physical security depends on his having a job: and his having a job depends on his taking instructions.
(e) Man's sole interest is the welfare of himself and his family.

All of those assumptions would now seem to be subject to challenge, no matter how ironclad they may have been in the past. Which is not to say, of course, that they are now totally untrue: to claim that would be to fly in the face of most of the evidence, and much informed opinion. What does seem to be true is that their validity has lessened: to assess by how much, each of them needs to be examined on its own.

(a) People will do as they are told by those in authority

Even in Adam Smith's day, with the American War of Independence followed swiftly by the French Revolution, and with mutinies a recurrent feature in most armies and navies, no-one would have claimed that this proposition was universally true. Nevertheless, in general it did apply. Centuries in which the decisions of authority had been promptly backed by physical force ingrained in most people a respect that more or less automatically resulted in the carrying out orders.

This respect was reinforced by the prevailing attitude of the churches. While the Protestant churches might owe their existence to having challenged the authority of Rome, with very few exceptions they continued to maintain that interpretation of the Commandment to 'honour thy Father and Mother' which extends it to mean 'obey those set in authority over you'—whether the setting is done by the Church or the State.

And, in the societies of Western Europe in the eighteenth and nineteenth centuries this was backed up still further by the respect for parental authority instilled in children, both at home and in school.

All of these factors have changed dramatically. Society at large has moved and is still moving in the direction of greater permissiveness, and not just in the narrowly sexual sense in which the phrase is generally used. In the face of that movement, the churches have retreated: it has become a relative and newsworthy rarity for even the Catholic church to issue unambiguous instructions, rather than advice, to its members.

And, perhaps, above all, in the home and in school attitudes to children have almost entirely reversed. For good or ill, we have for more than a generation been educating children to expect to participate in the decisions that affect them, and to challenge and question authority, rather than accept it automatically.

(b) People will accept that their superiors know better than they do

Education has played its part in the shift of attitude on this count too. For, while one may argue about the quality of education today as opposed to a generation or two ago, it is undeniable that the differential in education level between manager and managed has narrowed. In Sweden, which is not so far ahead of other countries, it is estimated that, by the end of the 'seventies, 90% of children will receive a highschool education. Once upon a time, there was some justification for the worker accepting that the manager probably knew better because he was better educated. Now that the gap is narrower, there is less reason for him to make that assumption.

Importantly, too, in a fast-changing world, there is less reason to accept that length of experience is a criterion of relevant knowledge: where promotion is largely based on length of service, this becomes another reason for reluctance to accept the better judgement of those in authority.

For both these reasons, the expectation of participation mentioned earlier is bolstered by confidence in the ability to participate usefully.

It is also worth noting here the growing readiness of workers to criticize managerial decisions, not on the traditional grounds that they are against the workers' interest, but simply because they are incompetent and inadequate. This particular phenomenon has probably been most manifest in the situations where workers have fought, by 'take-overs' or 'work-ins', the closure of a plant: but it is increasingly common in less extreme situations, and has probably been fostered by the crisis of self confidence which management in general has appeared to be going through in most countries in the last decade.

And it adds to the expectation of participation and the confidence in ability to participate the further feeling that it is *necessary* for the employee to participate in order to combat inefficiency.

(c) Man's prime motivation is economic gain

This is an area in which it is very easy to get into difficult and disputable waters. William Winpisinger, general vice-president of America's International Association of Machinists, holds out for the traditional point of view, putting it strongly and succinctly: 'The better the wage, the greater the job satisfaction. There is no better cure for the blue-collar blues.'

In any case, whatever the true ranking of human motivators, it is unlikely that they have fundamentally changed in only a century or two. The argument about the place of monetary reward in that ranking is hardly worth getting into, since one thing is undeniable: money becomes of less and less utility as one acquires more and more of it—as living standards and real incomes increase.

There is, therefore, no need to claim that there has been any change in human psychology. Money may still be important, and may even be getting more important. But the relative value as a motivator of any particular amount of money steadily decreases. It is still true that people can be induced to do any job if the money rewards are high enough: but the premiums

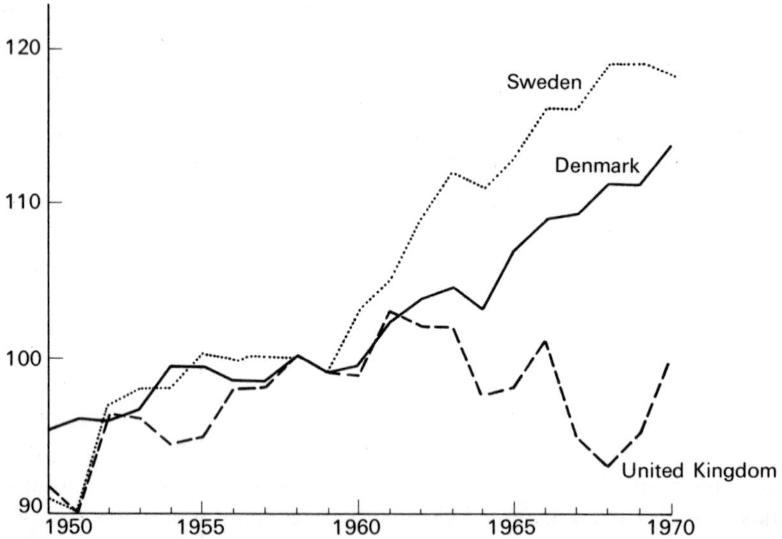

Fig. 1.3. The cost of buying productivity. Labour costs per unit output at constant prices. Index, 1958 = 100.

needed get greater all the time. Reliance on financial rewards as motivators becomes, therefore, steadily less profitable.

In this connection, Fig. 1.3 is interesting. The fact that emphasis, in studies of productivity, is usually placed on output per head tends to obscure the fact, but there has been no great improvement in the last few years in the labour cost per unit of output. In some countries, the cost has, in fact, steadily risen; in others, the increase in wages demanded has fairly neatly cancelled out increases in individual productivity.

One point about this chart needs commenting on. It may come as an initial surprise that the UK has done so much 'better' on this criterion than Sweden, for instance. If labour costs were the whole story, one would expect UK goods to have become much more competitive than Swedish goods. They have not, because the UK's record in controlling other costs is much worse.

(d) Man's physical security depends on his having a job: and his having a job depends on his taking instructions

To begin with, the second part of this assumption is visibly less true than it was. In an ever-growing area of industry it has become much harder, not to say more expensive, to dismiss a worker without due cause. And the definition of due cause has become more limited. Refusal to accept an instruction is less and less often considered sufficient.

The most recent legal provisions in the area, the Swedish law which took effect in July 1974, go so far as to make the only cause for dismissal either

— a *proven* lack of work
— *repeated and clearly documented* unsuitability

And not, in particular, 'single cases of neglected duties'. Even in either of the two accepted cases, six months' notice must be given.

If the union does not accept the case for dismissal, the case is heard in court. If that decision goes against the employer and he still refuses to take the man back, penalties, depending on age and length of service, can range up to four years' salary.

But the first part of the assertion is also less credible than it was. Objectively, it may still have something in it. Though most countries ensure that unemployment poses no critical threat to physical security, it usually leads at least to discomfort. But what counts is not objective security, but the *sense* of security. *Fear* of unemployment, and particularly prolonged unemployment, has diminished considerably in all the countries we are concerned with. (On the shopfloor, at least. In management, and especially at higher levels, insecurity remains a powerful motivator. It may even have increased in strength.)

One way in which this change manifests itself is that it is no longer any great cause of satisfaction—as it once was—just to *have* a job. For the satisfaction of status needs, what now counts is the *kind* of job. But the most evident way in which it has become apparent is probably that the link between unemployment levels and wage increases has been snapped. It used to be a fundamental principle of macro-economics that high unemployment levels were associated with low rates of wage increase, and vice versa. But in recent years, in most countries, mounting unemployment has signally failed to act as a check on wages.

(e) Man's sole interest is the welfare of himself and his family

Of all the assumptions, this is perhaps the most tendentious one, and its criticism perhaps the quickest way of arousing scepticism, and even ridicule. Yet, especially in America, commentators are more and more ready to assert (for instance, Samuel A. Culbert and James M. Elden in the *Harvard Business Review* of November–December 1970) that young people in particular are increasingly committed to a social ethic rather than an individualistic one: that their drive now is less and less for individual, and more and more for communal achievement. Allegedly, the sense of belonging and contributing to the group and the community replaces as a source of satisfaction the sense of having 'made it' on one's own.

It is important to distinguish this kind of motivation from the traditional one of group, or class, solidarity. Loyalty to one's own work group, to one's union, to one's class, is and always has been important, but it is at least open to the charge of being at bottom self-interested—a simple extension of family loyalty. The allegation here is not of an increase in that kind of loyalty, but of a growing desire to be and feel of service to the community as a whole.

Unfortunately, perhaps, while it is true that this kind of assertion plays a major part in the slogans of protest and dissent, that it is yet having any great effect on the workplace is not so obvious. Again, therefore, it may be as well to avoid the argument, simply record the assertion, and accept that, if it is true, it will only increase the effect of the other changes.

What changes there have been in these five underlying assumptions of the traditional system may, in some cases, have been only marginal. It would certainly be false to claim that the changes in attitudes have affected all, or even most of, the workforce. Objective estimates of the proportion of workers unaffected by changing attitudes, at least in any major way, are in fact remarkably consistent. In a typical study of 3800 workers in the US, A. A. Imbermann found that roughly 80% were not alienated by assembly work, while the other 20% were. An English study reported by Ian Mandle and John Lawless produced exactly the same proportion.

(Estimates by people more directly involved can vary widely, however. During the writing of this chapter, one British union official estimated that about 10% of the members of his union were satisfied with 'button-pushing' while about 15–20% would like their jobs made even simpler. The remainder would like to take a 'broader interest' in their job. Another official from the same union felt that no more than 5% were actively interested in participation, or changing anything.)

But even if the proportion of workers seeking or interested in change form a relatively small minority, that minority can be important in several ways. In the first place, this is the active minority: the union leader who reckoned them to be only 5% of the workforce pointed out that it was the same 5%

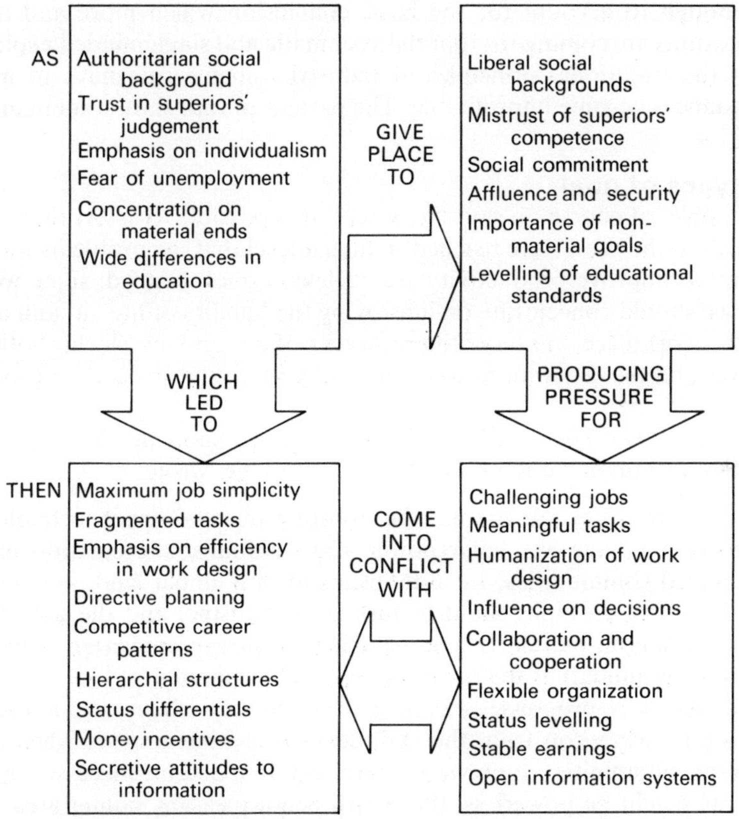

Fig. 1.4. The seeds of conflict.

who were also active in the union and in politics. They may well, therefore, be in a position to lead the majority.

Secondly, even if the changes in attitudes are taking place at the margin, in questions of increasing productivity it is the margin that is important (at least, once reasonably well planned production systems are already installed). Noncooperation from 20% of the workforce is quite enough to negate the value of most planned productivity improvements.

And, finally, we are talking, of course, in gross averages. In any particular firm, a national minority may easily be a majority. In a series of seven studies reported by Professor G. Rühl in Germany, 'satisfied' and 'dissatisfied' workers were evenly split. But the extremes were 75% dissatisfied and 16% dissatisfied: in two different departments of the same plant. Such concentrations of discontent, whatever their cause, are likely to be more important in practice than gross averages.

Marginal change or not, there appears to have been a shift in attitudes large enough to account for the basic conclusion which more and more organizations are coming to: that the systematic and singleminded exploitation of the traditional principles of industrial engineering have in many circumstances become unprofitable. The pattern of that shift is summarized in Fig. 1.4.

Two types of goal

It is possible, of course, to say: 'So what?' It is possible to assert that living standards in the West have reached so high a level that the problems we face in trying to improve productivity are irrelevant ones. Instead, some would argue, we should concentrate on improving the 'quality of life', at and away from the workplace, conserve the resources of a shrinking world, both for ourselves and the generations to come, and give up the struggle to produce more.

A letter to *The Times* in London 1974 perhaps sums up the argument: Mr Julian Cummins, of King's College, Cambridge, wrote:

> . . . The strategies are simple: the boosting of low-impact technologies based on methane gas, solar power and windmills; the creation of decentralized communities; the establishment of a global land, sea and air speed limit of 25 mph; the dismantling of industry; and the use of the bicycle (man-on-bicycle is a more efficient energy converter than any machine or animal) as the pivotal social and technological tool.
>
> This would be progress: away from the nightmare of recurrent industrial crises (and away too from the skill deficiencies that impoverished agricultural communities in previous centuries). In a decentralized society oil sheikhs would be powerless, the North Sea a pleasant sailing area, and nuclear power stations a figment of science fiction imagination.

Like most Utopias, such proposals find their followers. But we would not agree. This book is based on the premise that increasing productivity re-remains all important: indeed, that the pressure to increase productivity is greater than ever before.

There are many reasons. To begin with, there is no apparent limit to the pressure for improved living standards, no matter how affluent society becomes. The psychology underlying that pressure is not our concern: but it has to be registered as a fact: a fact that is, of course, even more true of the majority of the world than it is of the West.

Yet, at the same time, we are confronted with a shrinking in available resources—in energy and raw materials. If we count man-hours as a resource, then the shrinking is even more acute. For not only is the demand for leisure-time, both in holidays and shorter working weeks, growing everywhere, but in most areas population growth is decelerating, and in Western Europe there are recently signs that we may again be entering a period of decreasing population.

It is not the purpose of this book to dwell on either the quality of life or the need to preserve resources. But it seems undeniable that, if we are to meet the continuing demands for improvement in living standards (including in that phrase improved care of the sick, the aged and the unfortunate), and if we are to do that in the context of scarcer and more expensive resources, then the pressure to increase productive efficiency must go up. Unless the 'quality of life' is to be increased at the expense of living standards, it can only be paid for out of increased productivity.

That is from a world viewpoint: from the narrower point of view of the West, there are even stronger factors pointing in the same direction. One of Keynes's last predictions was that, after the 1939–45 war, the terms of trade would swing strongly against the developed economies in favour of the producers of food and raw materials. For a generation this did not happen (and the failure of the prediction was used by more than one economist as a propaganda point against Keynes). Finally, however, it seems that he is to be proved right: the relative value of the manufactured goods on which the West largely depends against the primary products of the rest of the world is dropping sharply.

To maintain its balance of trade the West must therefore get more output from its inputs: it must become more productive.

Moreover, if, in the West, traditional methods of improving productivity seem to have reached the point of diminishing returns, the same is not true of the immature industries of the rest of the world. Given the capital that, ultimately, exploitation of their bargaining position as suppliers of raw materials will bring, the countries of the Third World are, in many cases, poised to repeat the successes in industrialization of Japan and Hong Kong.

Already in the late 'sixties, the greatest improvements in productivity of any countries were being reported by Taiwan and South Korea; Persia, with the target of becoming the fifth most powerful economy in the world, is achieving growth rates of up to 40% a year.

They can do so simply by copying methods exploited successfully in the past by the West. But if the West is to compete, *it* will have to find *new* methods.

Against that background, the lesson for the individual company is stronger still. In such an environment, the company that survives will be the one that best succeeds in coping with the challenge of devising new methods of becoming ever more productive. The decisive competitive weapon for both companies and countries may turn out to be the ability to achieve and maintain cooperation towards that goal.

There are, of course, other arguments still. It may be passing a value judgement to say so, but the only society that can truly claim to be healthy is the one that is dynamic and growing. Admittedly that does not necessarily imply the traditional concept of growth: as the 'quality of life' advocates would suggest, there are, perhaps, more important and more relevant measures of growth than industrial output or GNP.

But—however it measures its growth, materialistically or spiritually, in quantity terms or quality terms—no matter what goals it sets itself, a society will only be able to achieve them if it can simultaneously achieve more effective and efficient ways of managing its material and human resources in the production of goods and services.

That, succinctly, is the challenge of the new chapter in the Industrial Revolution that confronts us, now that Adam Smith's old principle of the division of labour seems to have reached its limit: to find ways of uniting the 'old' economic values of the early Industrial Revolution with the 'new' human ones that have emerged into prominence, joining the two in new socio-technical production systems.

First thoughts on how that challenge may be met are the subject of the next chapter. For the moment, all that remains to be said, perhaps, is that, though our primary concern may be economic, it just may be that our economic concern may lead us some way to contributing to the goal for industry set by Pope John XXIII in his 1962 encyclical *Mater & Magistra*:

> . . . We believe . . . that the worker must be assured an active role in the functioning of the enterprise in which he is employed, whether it be public or private. We must attempt to make of the enterprise a true human community which marks profoundly with its spirit the relationships, functions and duties of each of its members.

2. The work and the worker

Two schools of thought. 'Scientific management': spotlight on the work. 'Behavioural science': spotlight on the worker. The need for synthesis.

In outlining in the rest of this book our own response to the challenge posed by the last chapter, we will be setting ourselves two limits. In the first place, our major concern will be with manufacturing industry. Occasionally, to make a point, we may venture outside that limit, but by and large we have preferred to avoid the complications that would be necessary were we to try and include the extractive and service industries, or even the highly automated process industries.

This must mean that, for people in those industries, some parts of this book (notably chapter 4) may seem irrelevant, or to have been given too much importance. It may also mean that we have omitted topics that would be of importance to those industries. But there is still a great deal of overlap: most of the central issues are common to all industry.

The second limit is that we have based our discussion on accepting the continuance of competitive, profit-motivated business in a mixed economy. It may be that we did not need to say that. But, certainly, one possible reaction to the problems of the last chapter is to blame them on the political system. For some people, the most convenient interpretation of dissatisfaction and conflict in the industrial system is ideological. In their view, alienation is the inevitable consequence of the profit motive.

Thus, Fritz Vilmar writes, in *Humanisierung und Demokratisierung der Arbeitswelt* ('The Humanization and Democratization of the World of Work'—our translation):

> People get excited (correctly) over the fact that men have to work in monotonous work situations. But when it comes to what for the worker is the most obvious and important point: that other people pocket the fruits of his labour, and arbitrarily decide his working conditions—then they either simply accept it as a fact of nature, or they treat it as a 'trades-union problem' which is basically no concern of social scientists. . . . This basic observation leads to the logical conclusion that strategies for humanization and democratization can in no way be limited to the sphere

of the plant or enterprise, without effective measures being taken at the same time to control the total economy, and to achieve equilibrium and social welfare through planning.

We do not agree. In the first place, there is no evidence that similar stresses would not be revealed in the State-controlled economies, were there freedom to express them. Indeed, the experience of Czechoslovakia in 1967–68 and Poland in 1970–71 both point to the existence of deep discontent. When Miklos Haraszti, a worker in the Red Star factory in Budapest, was sent to jail recently, his offence was that he had complained about the dehumanization of workers in his factory, striving to earn maximum income in a machine-paced system. While his fate was different, his message was much the same as that of industrial reformers in the West.

Here, too, that public ownership brings work satisfaction has to be proved. During the first General Election campaign of 1974 in the UK, Scottish members of the steelworkers' union formally protested against the nationalization of the industry, on the grounds that it bred alienation.

We therefore accept the profitability objective for the enterprise. And that necessarily means accepting a productivity objective of one kind or another for the factory. Whatever objectives it may turn out to be necessary to add to those two, they themselves cannot be lost sight of.

Two schools of thought

The history of attempts to improve productivity (and thence profitability) over the last several decades has been dominated by the existence of two schools of thought, which most of the time have been seen as being essentially conflicting. The first of them, which is generally viewed as stemming from the pioneering work of F. W. Taylor, we will be referring to as *scientific management*. The second school, which perhaps has its origins in the Hawthorne experiences of Mayo and Roethlisberger, we will be calling *behavioural science*. In both cases, the titles are not perfect. Both are designations that might well be rejected by some people that we would apply them to. For our purposes they should be treated simply as useful labels. Not too much meaning should be read into the phrases.

The first school has traditionally concerned itself primarily with the efficient design of work systems: with maximizing output and minimizing inputs and waste. Its approach has been mathematical and analytical, adopting as far as possible the attitude of the engineer or the technologist. Its concern, succinctly, has been with the job.

The second school, on the other hand, has taken as its primary concern the well-being, in one way or another, of the worker. Since 'well-being' is definable in many different ways, the school has included many different approaches. Some (at least in the eyes of their opponents) have been 'soft and woolly'. Others have been rigorously experimental. But in each case the focus of attention has been less the job, than the person doing it.

To some extent, this much-publicized division may well have its roots in the dual nature of man himself. Religion has to concern itself with both the body and the soul. Philosophy and medicine, using a different language, have to take into account both the body and the mind. All three subjects indicate, not only that two ways of dealing with man-at-work may be possible, but that both may be necessary. We may have to take into account both the physical–physiological dimension of scientific management and the emotion–psychological one of behavioural science.

But before we consider how that may be achieved, it is worth a short review of the kind of teaching and practice that we are including under each heading.

Scientific management: spotlight on the work

The core of the scientific management school, as we have said, is the efficient design of work systems. It depends on the systematic study and definition of work, its analysis into elements which can be critically evaluated, described, and put together again into expedient sets of tasks. It has tended to become a straightforward set of procedures for improving work methods, leading to a more favourable rate of exchange between input and output of work. As such, it has much in common with the approach used for improving machine design, or for measuring the financial efficiency of enterprises. Not surprisingly, therefore, its techniques have obvious appeal to engineers and economists.

From the late nineteenth century onwards, scientific management developed in a reasonably linear fashion. Its development was not always at a steady rate, since it depended on a series of brilliant insights, each of which produced a burst of accelerated achievement. But each improvement was essentially a move in the same direction.

After Taylor came Fayol, Gantt, the Gilbreths, Maynard, Stegemerten, Schwab: in each case their original contributions were successively applied, refined and enlarged by a growing profession of industrial engineers. But the line of development remained constantly directed to the same goals.

Perhaps the greatest common factor was the school's virtually uniform obedience to the maxim that maximum productivity is obtained when 'each worker knows what to do, how to do it, and in what time'. Ever closer and more objective specification of each of these parameters became one hallmark of the profession's advance.

For the industrial engineer, another hallmark of that advance is the finding of ever more general measures that apply to more and more situations.

We can, perhaps, obtain a deeper understanding of the approach by consideration of one reasonably typical measurement formula (which we will be returning to later in the book). This considers productivity as the product of three factors:

Productivity = Utilization × Methods × Performance

in which the three key-words on the right stand for:

Utilization	the ratio of the total time during which work is possible to the total time available (i.e., total time minus 'down-time' divided by total time).
Methods	for a particular task, the ratio of the time it would take using the best methods known to the time taken using the actual method.
Performance	the ratio of the worker's performance to some accepted norm.

Utilization is, therefore, largely a measure of organizational efficiency, including the flow of work, supply of materials, and availability of adequate machines and tools. *Methods* is a measure of the efficiency of task design. And *performance* is a measure of the worker's ability and effort. The way in which all three are defined demonstrates the overriding importance to the industrial engineer of being able to measure the current values of the factors in various situations, and to predict realistically the values that *could* be attained.

A numerical example may help to make the concepts clearer.

Let us assume a work situation in which the workers (for reasons outside their own control) are idle, as a matter of routine, 25% of the time. Utilization is, therefore, normally 0·75. Study of the situation and the development of possible improvements in the planning system, the supply of materials, or some other aspect of the organization, indicate that utilization could be raised to 0·85.

Moreover, due to various flaws, study reveals that the methods currently being used are only three quarters as efficient as the best that could be used. The method ratio is also 0·75. Which is to say that, for a given output, the best method known would need only 75% of the time required by the existing method.

Finally, let us also assume that worker performance, for whatever reason, is below what it could be. What should take 45 minutes to produce takes an hour. Again, the performance ratio is therefore 45/60 = 0·75.

The productivity of the system is therefore measurable as 0·75 × 0·75 × 0·75 = 0·42. The potential productivity would be, if all visible improvements were made, 0·85 × 1 × 1 = 0·85. Productivity would be more than doubled if the full potential could be realized.

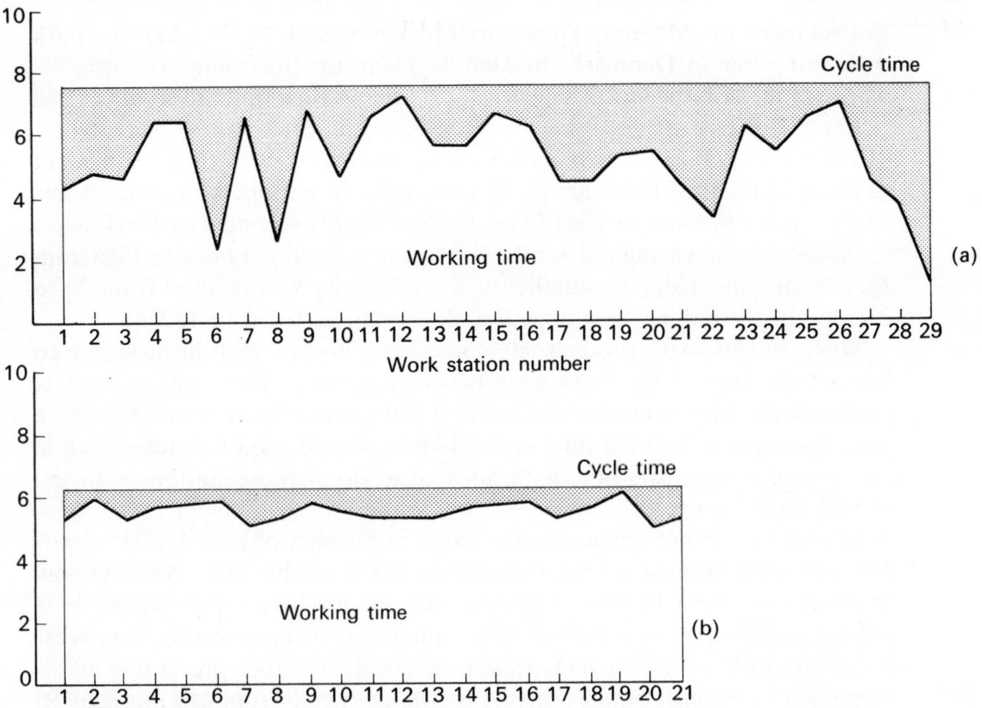

Fig. 2.1. The benefits of scientific management. Working time against idle time at each station on a pork process line.
(a) before study
(b) after line re-balanced.

Obviously, the effect on costs that depend on time (wages, machine running costs, interest and so on) would be considerable. If we assume that such costs are currently \$10 an hour and that ten units of output are produced per hour, time-dependent costs are at present exactly \$1 per unit. If the potential productivity were realized, unit costs would be reduced to a little under 50 cents, since the amount of output now producible in one hour would be $10 \times 0{\cdot}85/0{\cdot}42 = 20{\cdot}2$ units.

Realizing that potential, of course, might mean new investment in machines, and higher wages. But if such costs increased by 50%, going from \$10 to \$15 per hour, unit costs would still be only a little less than 75 cents—still a better than 25% improvement.

From approaches like these came the massive improvement in productivity in industrial society in the first two thirds of the twentieth century. The achievements of scientific management over this period are a matter of unimpeachable record. And it continues to report results.

Figure 2.1, for instance, shows graphically the results of one rationalization project using the Methods Time Measurement technique (MTM) on a pork processing line in Denmark. In each diagram, the horizontal axis lists the work stations on the line. The vertical axis shows time (in thousandths of an hour).

The horizontal line across each diagram represents cycle time: equivalent to the working time necessary at the slowest work station. As a result of this study, cycle time was reduced from ·0075 hours (27 seconds) to ·0064 hours (23 seconds), increasing the number of pigs processed per hour to 156 (from 133). Simultaneously, the number of work stations was reduced from 29 to 21: pigs processed per man-hour therefore went up from 4·6 to 7·4.

Much of this extra productivity came from improved utilization due to better balancing of the work time between stations. This improvement is shown in the diagram by the shaded and white areas. In each case the white area shows the amount of time actually spent working at each station, while the shaded area shows time with nothing to do at those stations with the shorter working cycles.

In a similar programme, Scania Vabis in Sweden obtained a 35% productivity improvement on its engine and truck assembly lines. And this was in an environment that had already been made fairly sophisticated by a previous study five years before. In a comparable defence project, one West European military maintenance shop managed to reduce preventive maintenance on a highly complex missile guidance system from 250 hours to 50 hours per unit.

There are abundant such examples. To quote a name more familiar, perhaps, in another context, Volvo of Sweden in 1956 designed a new factory for the PV 444 engine. Industrial engineers reduced the manpower required from 690 to 173. And Volvo has itself offered the cost/benefit estimate shown in Fig. 2.2 (from *Ekonomen*, No 8, 1957) of the three main factors contributing to the project:

Factor	*Cost Skr ('000)*	*Savings (No. of workers)*	*Cost per worker saved Skr ('000)*
Investment in buildings, machine tools, conveyors, etc.	11·3	185	61
Investment in tools, jigs, fixtures, etc.	2·5	82	30
Methods engineering, job design	1·2	173	7

Fig. 2.2.

Peter Drucker summed up the effect of scientific management in his book *The Practice of Management*:

> It may well be the most powerful as well as the most lasting contribution America has made to Western thought since the Federalist papers. As long as industrial society continues, we shall never lose again the insight that human work can be studied systematically, can be analysed, can be improved by work on its elementary parts.

As a recognition of the laurels due to the school there is nothing one can add to that statement. However, as we have pointed out, it had its rivals. It is time we turned our attention to them.

Behavioural science: spotlight on the worker

In contrast with scientific management, there is no direct linear development in the growth of the behavioural science school—or rather the group of assorted theorists and practitioners we are loosely lumping together under that title. Instead of a process of increasing elaboration of analysis and sophistication of methods, leading to increasing generality and scope, behavioural science confronts us with a series of competing and sometimes contradictory theories, with occasional and not always successful attempts at a synthesis.

They do have one thing in common, however. And that is their concentration on one of the three productivity factors we mentioned earlier: their concentration on *performance*. Indeed, their attention is focused on only part of that one factor, since their concern is essentially with the *effort* made by the worker, which is only part of his performance. (It is also affected by his innate physical and mental capabilities, his education and training, and so on.)

Their successive diagnoses of what affects that effort vary considerably, however.

In the earliest stages, the dominant thesis was that the major determinant of the worker's effort was the concern and care shown for him as a person by the enterprise. (In contrast, the scientific management school, then and now, held that the major determinant was how much he earned for it.)

The principal task for the enterprise that wanted to maximize effort, and thus productivity, became, therefore, the creation and maintenance of a benevolent and supportive working climate. As much as anything, this appeared to be based on the belief that, since work was necessarily unpleasant and undesirable, the physical and social atmosphere within which it was carried on ought to be as pleasant as possible. Present, too, was the belief that, if the enterprise demonstrably cared for the worker, then the worker would care for the enterprise, and for its results.

The logical appeal of that rationale, rather than any actual evidence of its truth, was enough to ensure a worldwide growth in interest in the various methods whereby such an improvement in 'human relations' in the firm could be achieved. The full range of such approaches became impressive, though what remained at the heart of them all was the development of welfare-centred personnel policies (in the UK, the Industrial Welfare Society became an influential pressure group—though later it was significantly to change its name to simply the Industrial Society). Other techniques on which emphasis was placed included company magazines and newspapers, competitions, and suggestion schemes.

On a more sophisticated level, there was also a vast increase in interest in management 'style', and in the development of training programmes to turn 'insensitive' (therefore supposedly autocratic and less effective) managers into 'sensitive' (and supposedly more effective) ones.

On utilitarian grounds, anything that makes people happier must be good. But it would be difficult to claim that its effect on the results of the enterprises involved was proven.

Above all, while it was usually fairly easy for the supporter of scientific management to point to the specific results of improving, say, work flow through the plant, it was not easy for the proponents of 'human relations' to do the same. It was inherent in their case that improvements would be long-term and indirect, and therefore difficult to demonstrate. While attempts to measure attitudinal changes were made, the connection with economic results remained intangible. Much had to be accepted on trust in the logic of the argument.

Unfortunately, the logic itself had flaws. In the first place, while the majority of people may be expected to respond positively to an expression of genuine care for them, there is little reason to expect the same response when the expression of care is felt to be merely the application of learned behaviour. Moreover, frequently the techniques involved were interpreted by some workers as an indication that management had 'gone soft'. And that only heightened any inclination they already had to exploit the situation.

Similarly, it was overlooked that the charisma of the autocratic leader can frequently overcome any resentment of his behaviour, at least in many social groups. The tough, 'insensitive', autocratic leader can frequently induce greater efforts than the one who is, apparently, eager to please.

And, finally, many of the well-intentioned attempts were seen by workers as 'paternalism', with all the adverse reactions that the term implies. Concern for employee welfare may have done something to lessen the frequency of industrial disputes. But when trouble did break out in 'paternalistic' firms (as recently, at the Pilkington's glass works in England) feelings were frequently more bitter than in many more impersonal situations. And on one side, of course, they were made worse by the employer's understandable reaction: 'After all we've done for them. . . .'

Ultimately, therefore, much of the work done at this stage (and still being done in the same vein) appears superficial and hollow. 'Cosmetic' would, perhaps, be a good term for it. But, as the behavioural science school became more mature (and more scientific) some of these obvious failings diminished.

The beginning of this second stage in the development of the school is perhaps best connected with the name of Abraham Maslow. Maslow developed (in general psychology, originally; not particularly in the industrial field) the theory that human needs could be ranked in a hierarchy. At the bottom come the physiological needs for those things that are essential for survival—air, food, drink. A little higher comes the need for security: above that the need for love and companionship: then the needs for status and esteem; and, finally, at the top, the need for self-actualization—the need to utilize one's abilities and talents, to 'be oneself'. His key proposition became that, while lower-order needs remain unsatisfied, higher-order ones are of less importance, and, conversely, once lower-order needs are satisfied, the higher ones come more and more into play.

Maslow's thesis pointed to factors that had previously been overlooked by both the devotees of scientific management and those of human relations. It was also important, however, in that it effectively denied the possibility of a permanent equilibrium. The satisfaction of one need only brings forth another: dissatisfaction is therefore a permanent feature of any human situation.

Indeed, dissatisfaction, on this view, becomes the motor of human progress, when handled properly. Just as, handled wrongly, it can generate destruction. One needs to prepare, not for a static situation, or even for a shift from one static situation to another, but rather for an endless journey. The only chance of stability lies in the stability of the rate of change (and its direction).

There were weaknesses in Maslow's formulation. Examples abound, for instance, of individual behaviour which seems to deny the existence of any fixed hierarchy of needs. For some people, self-actualization is more important than security, and even, perhaps, more important than eating. And there are enough martyrs in history to prove that staying alive is not always the most important objective.

Other psychologists, therefore, proffered alternative analyses of needs and goals. McLelland, for instance, placed major emphasis, at least in the industrial situation, on the need for achievement. More or less all of them criticized scientific management for producing work and organizational situations that overlooked or even went against significant human needs, and which were therefore counterproductive.

Douglas McGregor suggested that the fault lay in a singleminded attachment to one particular set of assumptions concerning people and work. Under what he called 'Theory X', people were assumed to dislike work, and to need coercion and control ('authoritarianism') or bribery ('human relations') to induce them to do it. They need to be directed towards the organization's goals.

McGregor pointed out that other assumptions were at least possible, and were perhaps more likely to be true. As an example, he invented 'Theory Y', in which people are assumed to have an intrinsic desire for work. By being given autonomy and responsibility and the freedom to be creative, they could be expected to contribute on their own to the organization's goals.

McGregor's point, contrary to widespread misunderstanding, was not that Theory Y was right and Theory X wrong. It was that there were many possible such sets of assumptions, any of which might be true in specific situations. It was up to the manager to clarify and test his own assumptions and develop a logical pattern of behaviour and organization from them. The most productive organization is the one that best fits the people and the tasks involved.

This last conclusion leads into a further stage in the maturation of the school, signposted in perhaps the most clear-cut way by Frederick Herzberg. Herzberg proposed that the only factors that could positively impel the worker to improve his own performance were those associated with the work itself: the challenge, opportunity for self-development, achievement and recognition that it provided.

Everything else—the financial rewards on which scientific management had relied, the climatic and environmental factors that human relations had emphasized, even many of the needs which the second-stage behavioural scientists had put forward, like security and the desire for group membership—were all peripheral to the work, and could neither provide satisfaction, nor impel the worker to work harder. If they were inadequate, they could create dissatisfaction, and therefore put a block on performance.

Herzberg derived from this proposition the further conclusion that motivation is latent within the worker. He firmly insisted that it is dangerously misleading to talk of 'motivating the worker'. All one can do is channel his existing motivation, tapping it by building as much opportunity for satisfaction as possible into the job itself, thereby 'enriching' it.

At this point our brief survey of the development of the two schools of thought ends. In one way, it may seem that we have come full circle. In leaving scientific management for behavioural science we left considering the work itself to concentrate on the worker. Now, with Herzberg, we appear to be focusing on the work again.

Unfortunately, however, in pursuing their separate paths around the circle, the two schools have not ended up facing in the same direction. Instead they have ended up in a face-to-face confrontation. One looks at the work from the point of view of how it contributes to the efficiency of production. The other looks at it from the point of view of how it contributes to the satisfaction of the worker. To the management scientist his opponent is soft and inefficient. The behavioural scientist for his part sees his opponent as callous and inhuman.

Instead of accepting that confrontation as inevitable, it is perhaps natural and obvious to attempt some kind of synthesis. In the past, proponents of both schools of industrial thought have been taking a one-sided view of the problems of work design. What one ought to be doing is seeking ways of using the insights of each school to repair the weaknesses of the other.

And that, in general, is what the rest of this book is about.

3. Designing production systems

A handful of fallacies. Four phases of design. Six sub-systems. Objectives for the system. Tangible/intangible. Economic/human.

From the way in which we ended the last chapter, it may seem that our recommendations in any practical situation would be an agreeable compromise between the two traditional camps. That would be a misunderstanding. We do not want to compromise, but to synthesize: to optimize. At each step in the systematic process of designing a production system we want to search out the solutions that best satisfy the legitimate demands of both the worker and the enterprise. We want to take into account, to put it another way, both the *economic* objectives the system is set up to serve, and also the *human* objectives that any social system has an effect on.

A handful of fallacies

For this integrated way of approaching the design problem, the phrase we would like to use is 'socio-technical production system design'. While a little long, and perhaps clumsy, it serves to underline the fact that the production system is neither a technical system nor a social system, but inherently both. And it must be treated as such.

Such a merger of interests, unfortunately, makes the problem more complex: it certainly does nothing to simplify things. It is not very likely, therefore, that the answers are going to be simple.

Yet, in the rush to abandon traditional industrial systems that has become fashionable in some places in recent years, it has sometimes been made to appear that the solutions *are* simple. Mostly, such solutions manifest themselves as slogans, and that is how we will take them here (in a list roughly based, with due acknowledgement, on the writing of Professor Rühl for the REFA-Institute in Germany). It should be remembered that they are all, in our view and in Professor Rühl's, fallacious.

(a) *The assembly line is dead.* In spite of the criticisms, and in spite of the number of companies that have, reportedly, abandoned it, the assembly line will necessarily remain with us—although maybe in modified, more sophisticated forms.

(b) *Cycle time cannot be less than one minute.* There are no hard and fast rules. Maybe it is true that, within limits, the longer the cycle time the more satisfying the job, but there are many other factors involved. And plenty of exceptions.
(c) *The answer to alienation is letting the worker produce a complete product.* It may well be *an* answer, sometimes. But most of the time it cannot be done.
(d) *Everyone should plan his own work.* Have a hand in planning it, probably. Even control the planning, sometimes. But while cooperation remains necessary, so does coordination and the place for the planning specialist.
(e) *Autonomous groups will make foremen (and the whole hierarchy) redundant.* At the fringes, the principle of division of labour may be turning out to be counterproductive. But that *is* at the fringes. For the most part, division of labour remains essential. So therefore does coordination. And so, therefore, does some kind of managerial hierarchy.

It is unfortunate, in a way, that such simple prescriptions all, in practice, fail. Life would be much easier if all one's experience and advice could be summarized and simplified into a handful of succinct slogans. But it cannot be done, at least in the area we are concerned with. There are no Ten Commandments of production system design.

Instead, the whole problem has to be viewed as a set of interlocking and interrelated problems, each of which is itself multi-dimensional. Such complexity can only be tackled through analysis. Which means that we are going to have to try to break down the whole process of production system design into its component parts in order to produce chunks of the problem that are small enough and cohesive enough to be dealt with on their own—while at the same time remembering that they are interrelated.

We can start the analysis by dividing the production system design process into four phases:

(a) The *survey* phase: in which the desired capacity and capability of the desired system are defined.
(b) The *resource* phase: in which the overall relationships between the human, capital and system resources are defined. Or, in computer terms, in which the roles to be played by hardware and software are established.
(c) The *system* phase: in which the interrelationship between the distinct sub-systems of the production system are defined.
(d) The *sub-system* phase: in which detailed attention is given to each of the sub-systems in turn.

(We have called these 'phases' because they follow one another chronologically. But that does not mean that they do not or should not overlap.

Feedback from work on the later stages may well influence decisions on earlier ones.)

Survey phase

The first of the phases, the *survey* phase, is largely concerned with factors outside the system itself. Its main concern is with the objectives the system will have to meet, and the resources available that it can draw upon. It takes into consideration, for instance, the demands made on the system by market factors; the limits that are set on it by financial, technological, or geographic constraints; its place in the company's strategic plan, if any.

In any specific situation, the results of the survey phase may well set constraints that severely limit the possibilities open in the following phases. Purely geographical constraints, for instance, may well limit the size and the kind of production system that can be installed in a shipyard. The width, shape, and flow of the river are not things it is easy to do much about. And of themselves they partially determine the way in which ships can be built on it.

Financial constraints may rule out the development or purchase of new equipment. Technological limitations may mean that new equipment cannot be developed, irrespective of the amount of finance available. Limitations on the amount of qualified manpower available may mean that tasks have to be of a certain degree of simplicity. And so on.

This means that, in any practical situation, work in the following stages is going to be easier than we may make it look. For, working in the abstract, we have to assume that all possibilities remain open. We have to assume in an assembly situation, for instance, that the designer is faced with a choice between a range of different kinds of assembly line, or even not having an assembly line at all. In practice there will rarely be any such wide choice: in most situations the range of possibilities is set, not by human decision, but by relatively implacable facts.

That does not absolve us from looking at all the possibilities. And, over a series of assignments, the production system designer may well have to be aware of them all. But it does mean that the problems are likely to be much simpler in practice than in theory. And in going through the complexities of the next few chapters, that ought to be remembered.

Resource phase

One way of approaching the *resource* phase is illustrated by Fig. 3.1. This divides the resources of the system into three classes: *capital*, *human* and *system* resources. The first two probably require no explanation. 'System resources', however, is a less familiar phrase. Effectively, we are using it to refer to the administrative methods by which all inputs and outputs of the

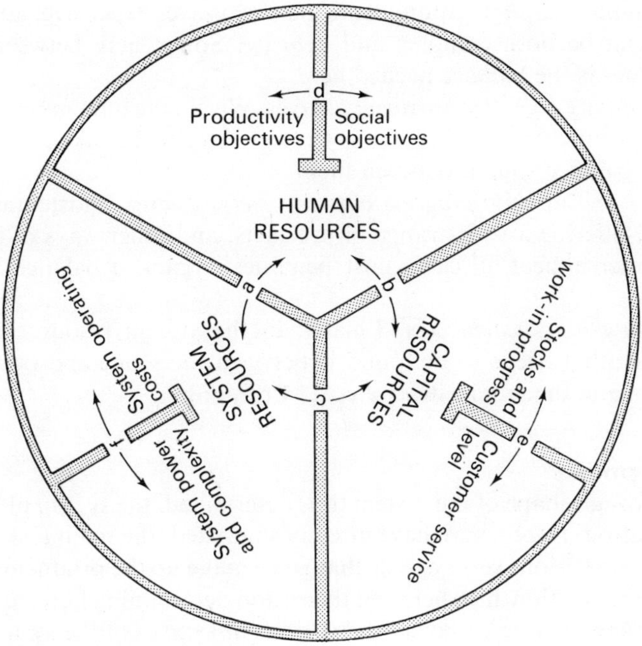

Fig. 3.1. The resource phase.

production system are controlled by volume, time, progress, quality, etc., and by which work is scheduled and routed through the plant. In almost any modern business situation, the costs of such administration are not only high, but continually increasing.

Three sets of resources means three sets of relationships (a, b, and c in Fig. 3.1) on each of which a balance must be struck. Moreover, within each of the three resource areas there is a primary balance to be struck (d, e, and f in Fig. 3.1). Thus, in the resource phase the following six decisions have to be roughly settled:

(a) *Human resources/system resources.* The balance to be struck here is between 'man-managed' and 'system-managed' planning and control: that is, the degree to which the decision-making should be 'automated'—with the proviso that 'automated' does not necessarily mean 'computerized' but may also mean 'bureaucratized'.

(b) *Human resources/capital resources.* Here the problem is to find the optimum point of the range from fully manual work to fully mechanized work.

(c) *Capital resources/system resources.* Extra stocks and work-in-progress necessitated by simple planning systems carry a cost penalty. But systems to minimize capital requirements are themselves expensive, and in addition can be both complex and delicate. Somewhere between the two extremes is the balance needed here.

(d) *Productivity objectives/social objectives.* Within the human resource area, both productivity and social objectives have to be taken into account and a balance sought between them.

(e) *Stocks and work-in-progress/customer service level.* Shortening delivery times, offering a wider range of products, and other ways of increasing customer appeal all carry cost penalties. Again, a balance has to be struck.

(f) *System power/system cost.* Finally, in the system resources area, the balance that needs to be found is between the scope and power of the system and the administrative cost of running it.

System phase

With the rough shape of the system thus determined, the *system* phase comes into operation. Here, as we have already suggested, the main task is to look at each of the various sub-systems that go to make up the production system, checking the relationships between them, and determining how, in combination, they can operate to satisfy the various objectives of the system.

Six sub-systems

These relationships are illustrated by Fig. 3.2, which compares the effect of six distinct sub-systems on six potential objectives. The choice of these six objectives is explained in a moment. First, however, the sub-systems themselves probably require a few words of explanation.

(a) *Production mode.* This sub-system, which is more fully discussed in chapter 4, refers roughly to the subject-matter of such questions as: 'assembly line or groups?', 'machine-paced or man-paced?', 'function-based shops, or product-based?' and so on.

(b) *Job.* By *job* we mean here the specific task the worker is asked to carry out at any particular time. It is important to establish the difference between the individual tasks themselves (which we would here call jobs) and the combination of tasks assigned to a particular worker, which is also frequently referred to as his 'job'.

This latter sense is the way in which the word is used in such phrases as 'job enrichment' and 'job enlargement', both of which involve changing the combination of tasks the worker is asked to do. In this sense, the worker's 'job' is affected by the other sub-systems.

In the sense in which we are using it, however, to refer to a specific task or operation, the design of the job is determined by ergonomic and

ECONOMIC OBJECTIVES			Production sub-systems	HUMAN OBJECTIVES		
Utilization ratio	Method ratio	Performance ratio		Security	Identification	Self-actualization
●		○	Production mode		○	○
	●	○	Job			
●	○	○	Work planning and control		○	○
○		●	Compensation	●	○	○
○		●	Physical environment	○	○	
		○	Organizational environment		○	○

● definite traditional influence ○ plausible potential influence

Fig. 3.2. Relation of production sub-systems to selected objectives.

work-study considerations that have not changed greatly as a result of the factors this book is concerned with. 'Job design' in this restricted sense is therefore something we will not be devoting specific attention to.

(c) *Work planning and control.* As a phrase, this should be self-explanatory. At least, nothing esoteric is intended by it. It is the subject-matter of chapter 5.

(d) *Compensation systems.* Specifically, the methods of financial remuneration to be used, which are dealt with in chapter 6.

(e) *Physical environment.* The worker's physical surroundings (other than the actual workplace arrangement). They are dealt with in chapter 7.

(f) *Organizational environment.* The extent and the way in which the production personnel are involved in the higher decision processes of the enterprise (see chapter 8).

System objectives

It is inherent in our basic approach that the objectives against which each of these sub-systems has to be considered are of two kinds: the physical (which we may call 'economic', 'productivity-centred' or 'traditional') and the psychological (which we may also call 'human', 'motivation-centred' or 'new').

There are, of course, many such objectives, and many ways of classifying them. Some of them are *tangible*, in the sense that they can be measured, and some of them are *intangible*. They may be intangible either because they are inherently unmeasurable, or because we do not yet have techniques for measuring them. And there are intangible and tangible factors on each side of the physical/psychological divide.

Usually, the importance of the intangible factors, being unmeasurable, has to be taken on trust. They have to be allowed for because of the plausibility of the arguments that support them, rather than because they have been empirically demonstrated.

But sometimes the intangible factors have directly estimatable tangible effects. Thus in Sweden, Saab-Scania, in calculating the cost advantages of switching from assembly line production to an assembly group system (as described in the Appendix), made use of the following breakdown:

1. Personnel turnover.
2. Absenteeism.
3. Quality.
4. Balancing.
5. Need for instructors and reliefs.
6. Recruitment.
7. Public and personnel relations.
8. Attitudes and cooperation.
9. Assembly time.
10. Supply.
11. Assembly tools.
12. Space requirements.
13. Handling system.

In the case of items 9 through 13, cost effects were not only measurable, but fairly accurately predictable. This is perhaps the highest degree of 'tangibility'.

For items 3 through 6, costs were not predictable (at the time of introduction of the system) but were measurable as they occurred. To that extent, therefore, they were tangible.

Part of the cost of items 1 and 2 was also directly measurable, though the effect on the productivity of remaining workers of high turnover and absenteeism remains intangible. The position of item 7 is different again, since, while the costs are tangible, the effects are not, directly anyway. And 8, of course, remains intangible, in the sense that any estimates of its effect must be either indirect or subjective.

The list is, therefore, a mixed bag. But it illustrates the range of effects that need to be taken into account in the system phase of the design process, even if it also illustrates the complexity and lack of clarity involved. (This point is made even more strongly by the Kockums shipyard case in the Appendix.)

In an effort to clarify the complexity a little, for Fig. 3.2 we have chosen only six of the various objectives that might be considered. While they do not cover the complete range, they do illustrate the varying ways in which the different sub-systems relate to them.

On the left-hand 'physical' side of the figure, we have used again the three components of productivity described in the last chapter: utilization, methods and performance. On the right-hand 'psychological' side, we have also simplified the range into three: security, identification, and self-actualization, each of which needs some short explanation.

By *security* we mean to include the individual's desire for physical safety, as well as for financial security, stability of employment and income, and so on.

By *identification* we again refer to a variety of possible components, including the desire to form a significant part of a group, and to identify with and receive recognition for the results of one's own work. We are therefore putting together, rather loosely, identification with the *work*, with the *product*, with the *work group*, and with the *company*, all of which will need separate consideration in later stages.

And, finally, by *self-actualization* we mean the need for achievement, responsibility, influence, personal growth, and the utilization of skills and knowledge, all of which again will need separate treatment at various stages.

We have thus lumped together into three broad groups (rather brutally, some may feel) the variety of human objectives and needs which the schools of behavioural science have put forward at various times. In Maslow's terms, security and self-actualization are respectively the lowest and the highest of the non-physiological elements of the hierarchy of objectives. Identification covers everything in between. In the terms used by Herzberg, security covers a group of hygiene factors, and self-actualization a group of motivation factors, while identification includes both kinds.

There is not, therefore, much academic backing for our classification, but it does to us seem to provide a simple, easily understandable, and useful grouping.

With this in mind, Fig. 3.2 gives an indication of which of the sub-systems relate most closely to which of the objectives the production system designer is concerned with. It also, therefore, provides an overview of the kind of balancing process that he has to go through in conducting the system phase of design.

Sub-system phase

The survey phase, the resource phase and the system phase are all we intend to deal with in this chapter. For the final phase, the *sub-system* phase, consists of detailed design work on each of the individual sub-systems, and that is too complex for compression to be worth while. Instead, with one excep-

				Chapter 4: Production mode	Chapter 5: Planning system	Chapter 6: Pay system	Chapter 7: Physical environment	Chapter 8: Organizational environment
PHYSICAL	INPUT	Working capital	Stock level	✓	✓		✓	
			Work-in-progress	✓	✓		✓	
		Investment	Facilities	✓			✓	
			Equipment	✓			✓	
		Cost	Materials	✓	✓			
			Labour	✓	✓	✓		
	INPUT / OUTPUT	Cost efficiency	Machine utilization	✓	✓	✓	✓	
			Labour productivity	✓	✓	✓	✓	
			RoC/RoI	✓	✓	✓	✓	
	OUTPUT		Quantity	✓	✓			
			Quality	✓	✓			
	OUTPUT / TIME	Time efficiency	Output/time	✓	✓	✓	✓	
	TIME		Delivery time	✓	✓	✓	✓	
			Machine time	✓	✓	✓	✓	
			Set-up time	✓	✓	✓	✓	
PSYCHOLOGICAL	SECURITY		Physical safety				✓	
			Job security					✓
			Job stability	✓				
			Wage security			✓		
			Wage stability			✓		
			Capital accumulation			✓		
	SECURITY / IDENTIFICATION		Wage level			✓		
	IDENTIFICATION	Group membership	Co. level			✓		✓
			Plant level			✓		✓
			Work gp level	✓	✓	✓		
			Status in group	✓	✓	✓		✓
			Personal identity	✓	✓	✓		
			Product identification	✓				✓
	SELF-ACTUALIZATION		Skill level	✓	✓			
			Variety/complexity	✓	✓			✓
			Responsibility/autonomy	✓	✓			✓
			Development potential	✓	✓			✓

Fig. 3.3. Production system objectives.

tion, each of the separate sub-systems is given detailed consideration in one each of the next five chapters.

The one exception has already been hinted at in our list of the sub-systems. For the reasons we explained there, there is no chapter on job design. Insofar as the phrase means 'task design', then we have nothing to add to traditional expertise in the subject. And insofar as it means grouping a number of tasks into one 'job', then it is covered by the other chapters, in particular chapters 4 and 5.

The basic pattern we have followed is to give a chapter to each sub-system. Figure 3.2, therefore, also gives an indication of which objectives can be expected to be most important in which chapters. But Fig. 3.3 gives a more detailed exposition of the same thing. It also, on the psychological side, gives a summary of how our three main objective classifications are built up, and, on the physical side, it substitutes for utilization methods and performance a list of perhaps more familiar objectives, derived from standard concepts of input and output.

Briefly, in this chapter we have reviewed the work that needs to take place in the three earliest phases of production system design, before getting down to the 'nuts and bolts' of the problem: the detailed work involved in designing the sub-systems. But it may already be becoming obvious that in order to deal constructively, not just with the symptoms, but with the underlying causes of the discontent and alienation that appear to be afflicting our industrial society, we will need not less but more solid and creative industrial engineering.

That conclusion should be borne out even more strongly by the detailed analyses of the following chapters.

4. Choosing the production mode

Process, mass, batch and unit. Traditional assumptions. Overlooked possibilities. Production flow. Work orientation. Worker assignment.

'Production mode' is probably an unfamiliar phrase. In the last chapter we rather hurriedly defined it as the subject matter of questions like 'assembly line or groups?', but a little more explanation will probably be useful.

The traditional breakdown

Traditionally, what we are calling *production mode* has been seen as something that could be of four kinds. Production systems were classified as *process*, *mass*, *batch*, or *unit* type. It was obvious that some production systems fell between two of these groups, or overlapped them. But in general the view has been that there was a kind of scale against which all production systems could be measured, that 'process' and 'unit' production were at opposite ends of the scale, and mass production and batch production fell between them.

For the rough classification of types of industry, this classification undoubtedly has uses. A good example was provided by the work some years ago of Professor Joan Woodward of London. She studied a considerable number of UK firms, looking for correlations between organizational pattern and success. And only when she divided the firms up into these four classes did she find any.

For our purposes, however, concentration on this scale is an oversimplification. Moreover, it is one that can lead to significant error. Partly, this is because it confuses things that are essentially aspects of the market and of the product with things that are properly part of the production system. And, partly, it is because it invites the system designer to ignore many of the possibilities that are open to him. This he does whenever he assumes that certain aspects of the system are *inevitable* results of other aspects, rather than independent.

Thus, for instance, he may feel that choice of the 'mass production' mode necessarily implies setting up a mechanized assembly line, dividing the work into easily learned, short cycle time tasks, each performed by individuals in predetermined ways: the classic *Modern Times* formulation. Our case, however, is that none of these things necessarily implies the others. Each has to be weighed on its own merits. Because you have an assembly line, you do not automatically have to have simple, short tasks, and you do not have to rule out group working or autonomy.

Equally, faced with a traditional 'batch production' situation, the designer may automatically assume that he has to group his machines and operatives according to their function, put them in separate shops for drilling, welding, casting or whatever, and rely on a complex work planning system to ensure that each product takes the right route through the several shops.

It is not difficult to see why such automatic responses have arisen. There is a natural human tendency to allow successful solutions, derived to fit specific situations, to harden into conventional responses that are applied in situations that they may not fit. There is also a natural human tendency to prefer 'package' solutions. It is simpler to take the 'mass production' package off the shelf, so to speak, and apply it to the situation it is supposed to meet, rather than sit down and work out each time what aspects are suitable, and what are not.

So it becomes automatic, for instance, to think that in manufacturing cars one ought to do just what Ford did. (And there are some signs that it may, equally erroneously, become fashionable to think one should do what Volvo did.) And, by extension, it becomes just as automatic to accept that in the Ford situation—of high-volume, reasonably stable demand for a succession of identical or closely related products—you adopt the Ford solution—the assembly line and all that automatically goes with it.

For a long time, of course, following that system worked. And while it worked there was no pragmatic reason, perhaps, to question it. It might be arguable that better solutions could be reached in other ways, but, basically, given an adequate and working solution there is not a great deal of motivation to spend time developing theoretically better ones that may or may not work in practice.

In some industries and some countries it may still work. In highly automated process industries, for instance, the present package solutions (which are, in any case, fairly new) may still be satisfactory. And in underdeveloped countries the Ford model for industry may still be practical. But the *raison d'être* of this book is that in most industries in the West, the traditional packages do not work as well as they should, and it is therefore worth taking the packages apart to see what other possibilities, or combinations of possibilities, there are.

In some industries and in some countries, therefore, this chapter may well seem irrelevant to the practical problems. In most, however, it is critical.

Overlooked possibilities

Once one starts to take apart the traditional packages, three things become obvious. Two of them are, strictly speaking, only peripheral to the production system, but they are nevertheless worth mentioning briefly.

One is that it has too often been overlooked that the production system can, through properly planned stock-holding, be effectively decoupled from the demand situation. Holding stocks enables (at a cost) mass production systems to be utilized efficiently even in conditions of erratic demand, and equally makes it possible to meet stable ongoing demand with a batch production system. This is already, therefore, a pointer to one weakness in the automatic assumptions that high-volume stable demand means mass production, and low-volume erratic demand means batch production.

This avenue has been fairly well explored, so we will not dwell on it here.

The second point is that redesign of the product mix can make possible otherwise inappropriate production systems. Building a product range out of modular components, for instance (as in example 5 at the end of this chapter), instead of using different components for each variation, can be an effective way of rationalizing the production system.

Product design is a little outside the scope of this book, but it is worth pointing out that feedback should be possible between the production system designer and the product designer. Collaboration between the two is likely to produce far more efficient results than the simple presentation of a predesigned product to the system designer with the instruction: 'Work out how to make that.'

The other major result of looking more closely at the traditional packages is the realization that trying to classify systems along one scale, with process at one end and unit production at the other, is fundamentally inadequate. There is more than one dimension involved. In fact there are three components that need to be taken into account separately. We will be referring to them as *production flow*, *work orientation*, and *worker assignment*, and we can briefly define them as follows:

- Production flow: the way in which the product moves (if at all) during production.
- Work orientation: the way in which tasks are specified. Briefly, it may be on a method basis ('do this this way'); on a craft or skill basis ('paint this') with the method used left to the craftsman; or it may be on a product basis ('make such and such an item') again with the method left unspecified.
- Worker assignment: the way workers are assigned to tasks, with the important distinction being whether they are assigned in groups or as individuals.

But we will be able to consider these in more detail more effectively if we first of all introduce the objectives that we can hope to achieve through them.

The objectives

In the last chapter, we gave some indication of the variety of objectives that are potentially affected. As far as this chapter is concerned, that list can be considerably simplified. For we are now only concerned with those objectives that are affected by the production mode. And, more strongly, we need only concern ourselves with objectives that are differently affected by different kinds of production mode.

Starting with the economic objectives, it is first of all true (though perhaps not immediately obvious) that output objectives alone are irrelevant to the choice of production mode. Any kind of production mode, strictly speaking, can be designed and built to produce any size of output. Onc could make motor-cars by letting a single craftsman build the whole vehicle. And, given enough craftsmen, you could meet any amount of demand that way.

It would be incredibly inefficient, of course. And ridiculous though it may be, that example may make the point that it is not the size of the output that determines the choice of production mode, but the desired efficiency, which implies keeping within certain input constraints.

Our concern in this chapter is therefore with input and efficiency objectives. That still leaves a considerable possible list. Stock levels, work-in-progress levels, costs of equipment, machine utilization, labour productivity, and many others will all be relevant.

We can shorten the list, however, by making a few simple assumptions:

— Other things being equal, the simpler the job, the less expensive the quality of staff needed will be, and the more efficiently it can be carried out. This is, of course, not the same thing as saying the more efficiently it *will* be carried out. That involves other factors that we will be coming to later.

  While generally still valid, this assumption is sometimes called into question. It may be becoming true that the simpler the job the more you may have to pay people to do it. But, while there may be a trend in that direction, especially in Scandinavia and parts of the USA, it would be overstating the case to claim that it was as yet having any great effect.

— Similarly, the simpler the demands made on the planning and control systems, the more efficiently and cheaply the work will be carried out.

  Again, this is not the same as saying: 'the simpler the planning and control systems are'. That is a question for the next chapter. Our concern here is with the requirements the production mode places on the planning and control system.

- Finally, the more flexible the system is, the more efficiently and cheaply it will cope with changes, whether in type of product or in volume, and whether planned or unplanned.

Of course, there will be other economic constraints that the designer has to meet. Investment cost will be one. But, as we said earlier about output, in general, production modes of any *kind* can be fitted into any particular investment constraints.

Occasionally, this will not be true. In one of the examples at the end of the chapter, the constraint of available space limited the production mode possibilities. But there is no general relationship between space and any particular production mode.

On the basis of these assumptions, we can simplify the range of economic objectives down to three: *job simplicity*, planning and control simplicity—or, in short, *controllability*—and *flexibility*. (It may be interesting to relate these back to the utilization, methods and performance components of productivity mentioned in chapter 2. It is evident that job simplicity affects methods and performance, while controllability and flexibility affect utilization.)

Turning to the human side, we have already indicated in Fig. 3.2 that the security objective is irrelevant to the production mode. There is no way of relating any particular kind of production mode to either physical or financial security. It may well be true that a wrong *choice* of production mode may have a disastrous effect on the job security of everyone involved, but that is a different matter.

The identification objectives, in contrast, are deeply affected. For the purposes of this chapter, however, we cannot treat them as a unit. *Product*

ECONOMIC OBJECTIVES			Production mode component	HUMAN OBJECTIVES		
Flexibility	Controllability	Job simplicity		Self actualization	Product identification	Group identification
Strong	Strong		Production flow	Weak		
Strong		Strong	Work orientation	Strong	Strong	Weak
Strong	Strong	Strong	Worker assignment	Strong	Weak	Strong

Strong Weak

Fig. 4.1. Potential impact of production mode components.

identification and *group identification* are affected in different ways by the choice of production mode. We need to keep them separate, therefore.

The situation is different with the *self-actualization* objectives: they are all affected in much the same way, so we can continue to deal with them, for the time being, as a group.

So we have six summary objectives to concern ourselves with: three on the economic side and three on the human side. Figure 4.1 indicates the potential impact on these six objectives of the three components of the production mode we defined earlier. In discussing that impact, the simplest way would seem to be to take each component in turn, explaining along the way a little bit more about what we mean by the names we have given them.

Production flow

Production flow we defined as the way in which the product moves during production, and we qualified that by adding 'if at all', because one extreme kind of production flow occurs when the 'product' does not move at all. This is naturally the case with major construction projects, for instance, but even in traditional shipbuilding the product moves at least once, when the still-unfinished ship is launched. At the other extreme, in a mechanically driven assembly line, the product may be in continuous motion from start to finish. The commonest situation, however, is one in which its motion is discontinuous to some degree or other. It moves, is worked upon, moves again.

Of all the many elements going to make up the production system, production flow is the one over which the system designer in any particular situation is likely to have least control. Or, even if he has control, it is the one where his range of choice is likely to be most restricted by available technology and other unalterable conditions.

For instance, there would seem to be no sane way of making most industrial chemicals, or refining oil, other than on a continuous flow basis. But, such extremes apart, even if the designer's range of choice in any particular situation is a narrow one, it is important that he should be aware of what the range is. And frequently it is broader than has traditionally been assumed. With a little ingenuity, assembly lines of a kind can be introduced into machine shop situations: or existing assembly lines can have their continuity of flow broken up.

Reducing the amount the product moves does not, of course, necessarily reduce the amount of movement in the system. Keeping the product still means that workers, tools and equipment have to be brought to it. In fact, there would appear to be a generally valid principle that the more the product moves, the less people and equipment have to be moved. But which costs more, or takes more time, depends very much on the particular circumstances.

	Continuous	Discontinuous	Stationary
Flexibility	−	+	+
Controllability	+	−	+
Self-actualization	−	0	0

Fig. 4.2. Production flow.

It is possible, however, to point to some general effects of increasing or decreasing the production flow. They are summarized in Fig. 4.2, which adds to the input estimates of Fig. 4.1 an assessment of the *direction* of influence.

By and large, the tooling and the degree of engineering of workplaces necessary to set up a continuous flow line are bound to make it product-dependent. It therefore does not respond easily to changes in product. Nor does it easily respond to changes in volume of demand. To some extent, it is either working or it is not, and when it is, its output per hour is fairly rigid.

Partly, this rigidity can be countered in designing the line. Some flexibility can be built in. Thus, the Saab production lines are designed to operate at any one of a number of different volumes. It is fairly easy to switch from one to another.

But such precautions, virtually by definition, can only be taken to cope with foreseen variations, and only a limited number of those. Moreover, even a line with such built-in flexibility is still highly dependent on the presence of the right people at the right time and place with the right materials. If one piece of material or one person is missing, products will start coming up off the line unfinished. More extremely, things may start to pile up until everything stops.

Finally, the more continuous the flow, the more difficult it is to provide buffer stocks of part-finished products.

All of these things indicate that continuous flow scores heavily and negatively on the flexibility objective.

In contrast, at least when the continuous flow system is operating smoothly, it is much easier to plan and control. By and large, what you put in is pretty well bound to come out, after an easily predictable lapse of time, at the other end, suitably transformed. In a reasonably stable environment, therefore, the controllability objective biases the choice towards continuity of flow.

However, the other end of the scale, the stationary product poses more problems than the continuous line, depending on the amount of movement of men and materials that has to take place. But it is essentially simpler than many kinds of discontinuous flow.

In fact, it is somewhere between the two extremes, in something like the traditional job shop picture, that the strain placed on the planning and control systems is greatest. Here it is that any one product may have to follow a complex routing pattern between machines, and where there are a large number of different products the planning problems become immense.

In normal circumstances, such problems will be held to be the fault of the planning system, and much work may be done on the planning system in an attempt to eliminate them. The attempt may be successful. But the chances are that, after the expenditure of much time and money, the only result will be a more complex planning system, plus the same problems.

In such situations, it might well have been much better to pay more attention to the choice of production mode (and/or rationalization of the product range) as in the Fokker example described in the Appendix. In particular, the use of the techniques of group technology (about which we will have more to say in the next chapter) may be invaluable in smoothing out the production flow, and thus lessening the demands on the planning system.

Looking at the human objectives, the more continuous the flow, the greater the need for conformity in meeting it. So continuous flow limits the scope for autonomy, and is scored negatively for self-actualization. The other scores given, however, are zero, rather than positive. *Of themselves*, making the flow discontinuous, or keeping the product stationary makes no difference to self-actualization.

They are sometimes held to do so. But this is the result of the conventional assumption that tends to associate discontinuous flow and stationary products with product responsibility and group working. And these are types of work orientation and worker assignment that are in reality independent of the type of production flow.

Work orientation

We briefly defined *work orientation* as the way in which tasks are specified. That was a little loose, and, in particular, we did not mean the *degree* to which tasks are specified. While important, that is treated as part of the discussion of work planning in the next chapter.

Effectively, there seem to be three basic ways in which the worker's tasks may be specified:

— The method to be used and the operations that are to be carried out are specified. We call this *method-oriented.*
— The kind of worker (based on a skill or craft classification) to be used at any particular point is specified. This we call *craft-orientation.*

	Method	Craft	Product
Flexibility	−	+	−
Job simplicity	+	0	−
Self-actualization	−	+	+
Product identification	−	−	+
Group identification	0	+	+

Fig. 4.3. Work orientation.

— Only the end result (whether product or product component) of the operation or series of operations is specified. This we call *product-oriented.*

Since we have already mentioned group technology, it is worth noting here that its use can help to make the change from method to product orientation. The classification of products and components on which group technology depends can be used to identify product 'families'. In turn, these can be used to change a traditionally method-oriented job shop (with machines grouped by function) into one in which the machines are grouped according to the product families on which they are to be used.

Without the insight into product structure that group technology gives, such a change in the production mode may be impossible. In considering the effect on the objectives of different kinds of work orientation, Fig. 4.3 plays the same role that Fig. 4.2 did for production flow.

As far as flexibility is concerned, both method and product orientation score negatively. Method orientation makes it difficult to change the processes used. Product orientation makes it difficult to change the product line. In both cases a varying degree of job redesign and retraining will have to take place. The craft-oriented approach has none of these difficulties. Neither products nor methods are as permanent as crafts.

The method-centred approach has its chief advantage against the job simplicity objective. Maximum simplification will usually be attained when the operations to be carried out are specified by a job designer. On the other hand, product orientation virtually ensures that any worker's job is going to be a more complex one.

Craft-orientation lies in an intermediate position, for reasons that may not be obvious. The jobs the craftsman carries out may, objectively, be complex. But to the craftsman, thanks to his previous training, they are, effectively, simple. The costs of the apparent complexity are therefore not borne within the system (except insofar as the craftsman requires higher pay).

Turning to the human side, method specification scores negatively for self-actualization. It is probably readily apparent that method specification gives far fewer opportunities for job enrichment and autonomy. (It is, in fact, the association of method-specification with the traditional picture of the assembly line that gives it its currently bad image.)

Craft and product orientation both help provide job enrichment and autonomy, albeit in different ways.

With regard to product identification, product orientation, by definition almost, scores positively, whereas both the other methods of specifying tasks distract the worker's attention from the product.

It is, perhaps, surprising that we do not give a negative score to method orientation on the group identification objective. But it needs to be remembered that we are not talking necessarily of the specification of an *individual's* methods. We may be talking of a group: method orientation is quite compatible with group working.

Similarly, it is not because they are automatically associated with group working that we give positive scores to craft and product orientation. Both approaches can be used for individuals.

However, concentrating attention on an individual's craft status, even if he is working on his own, is bound to increase his sense of identification with other members of his craft. And concentrating attention on the end product necessarily increases the sense of membership of the organization involved in producing (and distributing and selling) it.

It is worth bearing in mind that, while craft orientation helps with group identification, it may not do so in a way that is beneficial to the enterprise. Membership of the craft may well become a stronger motivating factor than commitment to the organization. The result in some cases may be that the company feels it necessary to try and break it down, rather than build it up.

Worker assignment

We defined this component as being simply a matter of whether workers' tasks were assigned to them individually or in groups. To some extent, it does represent a simple division into two approaches. What is chiefly important is whether the worker is involved in his task purely as an individual or as a member of a group.

In most practical situations, the size of the group will have an effect, of course. The system designer will have a range of choices of which the effects will need careful evaluation. And it is possible to put forward a general

	Individual	Group
Flexibility	−	+
Controllability	−	+
Job simplicity	+	−
Self-actualization	−	+
Product identification	−	+
Group identification	−	+

Fig. 4.4. Worker assignment.

principle that groups larger than 15 or so should be avoided, since they are likely to involve both organizational and physical penalties.

But, from a theoretical point of view, the size of the group is less important than the basic distinction: whether or not the individual shares responsibility with others for an identifiable task or physical output.

What makes this last qualification necessary is that, to be successful, work groups have to be meaningful, and to be meaningful they must have some foundation in function or product. Arbitrarily designated 'groups' are likely to have no effect at all on the human objectives.

Calling a number of people a group simply because they work on the same product, or in the same machine shop, or sit on the same assembly line (or inhabit the same boardroom, for that matter) does not automatically generate group involvement. It can only do that if they are also held collectively responsible for a collective achievement. As long as they are held individually responsible for only their own activities, the situation is one of individual, not group, assignment.

Degrees of group assignment are possible, of course. For anybody's 'job' in the broad sense may include any number of 'jobs' in the narrow sense. For some, he may work as a part of a group, and for some he may work on his own. At the moment, however, we are concentrating on the task being performed at any particular time, not the variety of tasks a worker may be asked to undertake.

Figure 4.4 shows the direction of the impact of each of these two assignments on the relevant objectives. It is much more consistent than either of the two previous similar diagrams for production mode and work orientation. On all objectives save one assignment on a group basis scores positively and the individual basis scores negatively.

On flexibility, the superiority of group assignment derives from the interchangeability of group members (it may be worth repeating that we are talking about a collective group, not simply a collection of individual specialists). Interchangeability provides better protection against the effects of sickness, absenteeism, and other interruptions to people's availability. The task of planning and control is also simplified when all that needs to be taken account of is the aggregate of the group's activities. (We will be returning to this important principle of aggregation in the next chapter.)

As far as group identification is concerned, the superiority of group assignment is, perhaps, self-evident. The effect on product identification is also positive, though less marked, since group assignment inherently helps the possibility of adopting product-centred work orientation.

On the third human objective, self-actualization, the benefit of group assignment derives from the fact that the work of the group will normally be more complex. It will, therefore, normally offer more possibilities for enrichment and for self-development.

This same factor accounts for the one case in which individual assignment is superior: jobs for individuals will normally be simpler.

Opening up the general problem of the choice of production mode in this way, by separating out the different components and looking at their different effects, vastly increases the range of imaginable solutions to any production design situation. It certainly counteracts the tendency to rely automatically on one or other of the traditional packages.

Thus, when the three components are kept separate in the mind, it becomes possible, in a situation where economic considerations point to adopting a highly continuous flow system, to attempt to make up for the possible failings of the system on the human objectives by using a group approach to work assignment or a product-centred approach to work orientation. Similarly, the stresses on the planning system produced by a complex discontinuous situation in batch manufacturing can again be helped by shifting to group assignment, and perhaps also by changing from craft-centred to product-centred work orientation. And so on. The possibilities are many. Perhaps the one thing they all have in common is that the traditional skills and techniques of industrial engineering still have a considerable role to play in choosing among them. And, even more importantly, in producing the technical and technological advances which alone can make possible the blending of the three components in ways other than the traditional.

That said, we are in a position to take a look at a few examples from real life of production mode changes and their effect on some of the objectives we have been considering. None of the examples is a startling one. They have all been chosen as typical situations, not exceptional ones.

Example 1. Rubber boots

This first example, from a manufacturer of rubber boots in Denmark, dates back to 1958. As one might expect from that date, any human benefits resulting from the changes that were implemented were, to some extent, stumbled across by chance. Psychological factors were hardly dominant in people's minds in 1958.

The factory concerned had, in its day, been a progressive one. As long before as 1933, it had installed what was then a very sophisticated, machine-paced assembly line system. Twelve identical lines each occupied 18 men, with a basic cycle time of 45 seconds. At the time of installation, the system appeared to fit the situation extremely well. There were only two models of boot (urban and rural) in production. Demand was highly seasonal, being concentrated in the winter half-year but stock-holding costs were eliminated by the fact that labour was easily available on a seasonal basis. Workers were hired in August and fired again in April: the short cycle time made learning time unimportant.

By 1958, however, the situation had changed considerably. The range of models and sizes manufactured had multiplied, with the result that the lines frequently had to shift from one model to another. Moreover, the labour market had tightened to the extent that seasonal labour was effectively unobtainable. Workers had to be kept employed all the year round, which meant building and holding stocks. It also meant that there was no longer any particular advantage in having jobs with a short learning time.

These changes, however, were not the motivation for wishing to alter the production system. That came (as so often) from the availability of new technology: in this case, automation of the process whereby thin sheets of rubber were laminated to form the basic material for the boots. Introduction of the new machinery implied removing six men from each line of 18, and promised a return of over half a million Danish Kroner a year in return for an investment of three hundred thousand. But, at the very least, it meant rebalancing the line.

Theoretically, of course, that could have been done. Instead it was decided to evolve a system that fitted the demands of 1958 directly, rather than one that compromised on the solution of 1933.

Each 18-man assembly line was replaced by four short lines of three men apiece. Each line concentrated on longer runs of a more limited range of models. Cycle time rose from 45 seconds to 3–4 minutes. Theoretically, therefore, there was no increase in output: but in practice there was less down time for changing models. (And, of course, full advantage was now being taken of the new labour-saving machinery.)

It was only at this stage that some of the further benefits of the change became apparent. The old line had been machine-paced: and in any machine-paced line there is always likely to be conflict between slower and faster workers; slower workers will have trouble keeping up the pace and be stretched and insecure, while faster workers may feel frustrated at not being able to earn as much as they could working at their own pace. Even in unpaced lines faster workers are likely to feel held back by slower ones, while the latter may feel pressurized by their more highly skilled colleagues.

In the new system, with work groups of only three people, it became possible to team together the keener workers, reducing this particular cause of dissension. It also became easy to allow each of the groups to determine its own work pattern, and take its breaks whenever it felt like it. In the previous system, no matter how much responsibility for such things the management had been willing to delegate, it would have been effectively impossible to achieve the high degree of communication and group cohesion necessary to make such autonomy work in practice.

Example 2. Lorry axles

A current situation in Germany, involving the assembly of lorry axles, closely parallels the previous one, in that several short lines, each concentrating on particular product variants, again replace one long line.

The motive for change, directly, was once more an exogenous one—in this case it was the availability of extra space, thanks to the removal of part of the plant's operations to another site. This time, however, the human advantages of the smaller groups were borne in mind to begin with: but, as it happened, economic interests turned out to coincide.

Among the range of axle types being assembled on the original line, total work content varied between 60 minutes and 214 minutes, with 20 two-man stations; cycle times varied between 1·5 minutes and 5·4 minutes. Axles were produced according to a programme set by the total vehicle programme: the type required could change several times a day. Each time it changed, of course, the line could still only move (it was machine-paced) in accordance with the slower of the two products being assembled on it. At the extreme, workers could find themselves idle for virtually four out of five minutes while the change worked its way along the line.

By contrast, installing four lines, each specializing in one range of axle type, allowed the range of cycle times to be minimized considerably. The respective ranges became:

Line	*Stations*	*Cycle time range*
1	7	5·7–9·6
2	4	7·5–10·4
3	4	15·8–19·4
4	10	8·9–10·7

Additionally, of course, there were gains from the lower frequency of changes. And, moreover, the new shorter lines are not machine-paced (axles are on trolleys, mounted on rails, and pushed from one work-station to the next), and provision is made for small (one-two) buffer stocks between stations.

Costs for this kind of lack of balance were calculated for two- and three-line systems as well as the four-line system already mentioned. They are shown in the following table, which also includes the effect on labour and machine costs of holding different amounts of buffer stock at the end of the line.

	Annual costs (index) with a buffer of			
No. of lines	*1 day's output*	*2 days' output*	*3 days' output*	*4 days' output*
2	100·0	50·6	32·2	18·1
3	84·8	41·9	26·0	13·4
4	67·9	30·9	17·5	5·6

By the time the storage costs of holding stocks were taken into account, the optimum buffer size turned out to be two to three days' output, but the superiority of the four-line arrangement is evident. It was inferior in terms of space required, but not by a great deal: the three alternatives required respectively 402·5, 420, and 437·5 sq. metres. The slightly higher number of work stations meant a slightly higher demand for labour: but, again, the costs were relatively small compared to the gains.

The intangible human benefits are not so calculable: but at least the economically advantageous system coincides with that that should, by hypothesis, also be the more favourable for human reasons.

Examples 3 and 4. Cotton coats and wall switches

These two examples are grouped together because they both illustrate the same key point, though in different ways.

One is from a Swedish company manufacturing a variety of designs of cotton coats. At the outset, the production mode in force could be classified as discontinuous flow, craft centred, with individual assignment. Machines were placed in rows in the shop, and each of the several models in the range traced a different path from one machine to another. Apart from a rather low general level of efficiency, one specific disadvantage of the arrangement was the number of bundles of part-finished coats lying around the shop waiting for processing at one or another machine.

An essential precondition of improving this situation was rationalization of the product range, which turned out to be achievable without too much difficulty. Coats, it was determined, came in one of two distinct types: with set-in sleeves or raglan sleeves. Two basic patterns were therefore necessary: on each, some half-a-dozen points (collar, sleeves, pockets, for instance) were made design options. Order was thus imposed while maintaining a reasonable amount of design flexibility.

An assembly line was introduced. However, it had complications which make it worth noting.

In the first place, it was a suspension line: each coat hung from the moving conveyor on a separate hanger, eliminating the problems of bundling. Next and more importantly, the line had sidings, one at each workplace. While the line itself might, therefore, be continuously moving, at each workplace the coats stopped for an indeterminate length of time: so the work was self-paced. Moreover, each siding had room for a buffer of up to ten coats, giving even more freedom for flexible pacing.

Finally, each model of coat needed to visit only a certain selection of workplaces, in a certain order. As each girl finished work on a coat, she 'addressed' its hanger by pressing a button on it, put it back into the moving line, and it was automatically delivered to the next suitable workplace—as long as there was room in the buffer.

(To complete the picture, five off-line sub-stations did the necessary preparatory work on each design option: component parts were then fed up to the girls in the main line.)

From an output point of view, the innovation was a considerable success: the same number of girls produced 65% more coats, and work-in-progress levels went down substantially. Suddenly there were no bundles of coats all over the shop, which, apart from the improvement in work-in-progress produced the highly intangible but undoubtedly beneficial effect of a tidy workplace.

But, at least in hindsight, the most interesting aspect was the introduction of buffered sidings into the line; and the way in which control over work flow was exercised, through the addressing system, by the girls themselves.

The next example, from a Danish manufacturer of wall-switches, develops the same theme. Again, the major problem was making provision for frequent changes of model, and again, the solution involved a continuously moving line with sidings. There were, however, significant differences.

In the first place, each 'siding' involved a group of five or six girls (there were 80 on the whole line), not working together, but sitting together: work was on an individual basis, but there was plenty of opportunity for the formation of social groups. Girls in the same group might or might not be working on the same process, or even the same product: each had a separate table, which could be taken out and replaced with another fitted for another operation in two minutes.

The products and the components to be assembled were small. They were therefore supplied in boxes, which, to begin with, were stacked in racks at the end of the line. At her table each girl had, at any given time, a box of part-finished products and a box of components: she also had a set of buttons. As either box was filled (or emptied), she could press a button, and the requisite box would be taken from the automatic rack, addressed to her, and delivered to her station. In fact, of course, she would normally press the button shortly before finishing a box: the timing was up to her, as was also the pace at which she worked, at least within broad limits. The boxes of products on which she had finished her operation were replaced on the line, and delivered back to the racks, where they became input boxes for the next stage.

This last part of the process, the turning of output boxes into input boxes, was done by a man, not automatically: by doing it himself, the supervisor was easily able to keep an eye on the progress of each of the girls.

This system enabled one supervisor to control 80 operatives with no loss of efficiency. It also removed the need for manual supply of components, which had been necessary with the previous, more orthodox, line. This, to the designers' surprise, turned out to be especially popular with the girls, who, it transpired, had previously had to feed back part of their piecework earnings to their suppliers, in order to ensure an efficient supply.

But, again, the most significant point is the way in which reasonably imaginative engineering enabled most of the benefits of a continuous flow, individual assignment, method oriented production mode to be coupled with considerable provision for autonomy in pacing one's own work.

Example 5. Colour television

The final example for this section is also a current one, involving a Danish manufacturer of colour television sets. The major motivation for change here was straightforwardly the introduction of a new product line. But there was also pressure to reduce costs (to restore competition-pressed profit margins), and, importantly, to increase flexibility of output. The orthodox assembly line that had been in use for the old models was, according to the production manager, 'an elephant'. Designed to make 230 sets a day, it made 230 sets a day. To increase output appeared to be virtually impossible, and to decrease output, short of going to four or three-day working, equally impossible.

The key to developing a new system that would go some way towards eliminating some of these faults turned out to lie in the design of the new sets. Not for production reasons, but in order to ease customer service and maintenance, the sets were designed on the modular principle. Colour controls, sound controls, etc., were each built in a separate module that could easily be taken out and replaced. The decision was therefore made to treat each module, in the production process, as, effectively, a separate product.

There was also a second, separate, decision, facilitated by the first, which is more easily describable after considering the old system. Essentially, this was what we have described as a craft centred system (though within that mode, methods were fairly tightly prescribed). Components were assembled in one area, taken for machine soldering, then on for checking, then for repairing if necessary, then for hand soldering of heat-sensitive parts, then for more checking, adjustment, and so on. Each function was carried out in a different area by a different set of people.

This created problems, particularly in the quality control area. Feedback from testers and repairers to assemblers was necessarily slow, and could be nonexistent. One individual involved, exaggeratedly perhaps, reported that the people at the beginning of the line could go on making the same mistake 'for days', with people further up the line solemnly putting it right. What feedback was possible could only be through section supervisors; and like most supervisors they tended to be defensive about the work of their own sections.

The second decision, therefore, was to provide immediate feedback by letting workers do their own checking and repairing.

The system decided upon was roughly as follows. For each module, a separate line was set up, each manned by a group of no more than nine people. Because of the cost of the equipment, lines shared an automatic soldering station; but otherwise each line was responsible for the complete assembly, testing and final adjustment of its own particular module. After initial assembly, the module was sent by conveyor to the soldering machine: it was then brought back to the group for checking and all subsequent processes. Each line was provided with test-bed facilities for ensuring the module was in working order when it left the line.

Subsequently, there was an intermediate assembly stage in which modules were put together into sub-assemblies, which were then mounted into the final set: but much the same principles of assembly were followed at each stage.

Within each of the short lines, job rotation was encouraged: workers on the line were involved on a group basis, not an individual one. A group financial reward scheme has been put into effect. Higher-order MTM standard data were used as a common basis for methods engineering, workplace design, tooling and the pay system.

Some of the advantages are obvious. Quality control was immediately more efficient. Mistakes might be made, but they were now brought home very quickly directly to the people who had made them: no-one could go on any longer making the same mistake 'for days'.

Output per man hour was doubled; flow-through time was reduced from 20 days to 6; flexibility to meet capacity change was greatly increased and sensitivity to absence, material shortage, etc., similarly reduced. Work-in-progress went down, less transportation and handling damage was caused and overall production control was significantly simplified and improved by the aggregation of load centres and material supply.

Several concepts of job-enrichment and job-enlargement were successfully integrated with the work design.

In addition to forming 'semi-autonomous' work groups responsible for inter-group work allocation, volume and quality, there have been several other interesting advances. So, for example, it proved possible to increase the cycle time for some assembly operations from 15 to 40 minutes but only after the inherent work instruction problems were solved. This was achieved by the installation of an automatic projection unit which, at the touch of a pedal, throws a spot of light on the place where a specific component should be inserted on the printed circuit plate. Simultaneously the revolving component rack is brought into position so that only the correct component is immediately accessible.

All in all, a radically new assembly factory was created and run in. The past year's operating experience has proven beyond doubt its superior efficiency and attractiveness as a place to work in comparison to the one it replaced.

5. Planning the planning system

Animating the production mode. Delegation, participation, aggregation. The objectives. Six issues . . . and two techniques. The computer bogey.

The question of local and short-term work planning is the aspect of the whole planning system that most directly affects the worker. It is the work planning sub-system that determines how he shall spend his working days. And it is in planning his work that he can most obviously and most readily play an integral role. Here it is that, either individually or as a member of his work group, he can most easily make decisions, or, at least, influence the decisions that are made.

We make no apologies, therefore, for elevating to a prime position in the book what may seem a rather pedestrian subject. For work planning does not have the glamour or prestige associated with corporate planning, long-term planning, diversification planning, and such rather more fashionable topics. Generals are remembered (and have books written about them) and not subalterns and sergeants, let alone private soldiers.

It is true that, in one sense, work planning is less important than its more prestigious relatives. Individual errors do not make so much difference. No mistake in preparing a work schedule could ever have as much effect on a company's future as Curtis-Wright's decision in the 1950s not to develop jet engines. No mistake in controlling the flow of materials through a workshop could ever come as close to ruining a company as General Dynamic's failure to control its Convair 880 programme. And no single piece of work planning is ever likely to produce as much success as Rank's strategic decision to ally themselves with Xerox.

Yet, cumulatively, the difference is not so great. The continuous drain of an inefficient work planning system can bring a company low as effectively, if not as quickly, as an outstanding strategic error. More happily, superiority in planning and controlling its day-to-day operations may well be precisely the factor that allows a company to move ahead of rivals who are attempting to achieve the same strategic targets.

While this chapter may be somewhat unglamorous, it *is* nevertheless important. The elements of any work planning sub-system will fall essentially into one or other of two major classes, which we may call:

— Output planning (determining what quantity shall be produced of what products in what time-period).
— Input planning (determine how that quantity will be produced with the use of what resources of men, machines and material).

In some industries, of course, there may not be much freedom of choice in either or both of these classes. In automated systems, inputs required may be fully determined by the output required. In one-product process industries working to capacity there may also be no freedom in making output decisions.

Where such conditions apply, the work planning problem virtually vanishes. It may still exist for maintenance and repair staffs but, as far as the production system proper is concerned, it shrinks almost to nothing. It does not vanish altogether. Staffing levels have to be maintained, raw materials have to arrive on time, and control systems still have to operate. But, by and large, as we pointed out in the last chapter, some production modes place far less strain on planning and control systems than others.

Three key factors

In the general approach to designing systems for making work planning decisions, a mistake is frequently made which is closely related to the mistake we have already discussed, of assuming that we are restricted to certain 'packages' in choosing the production mode. In work planning, the equivalent mistake is to assume that the basic problem is settling on a position in a scale that runs from centralization to decentralization.

In the production mode, the mistake results from confusing production flow, work orientation, and worker assignment. In the work planning system (and elsewhere, as we shall be describing later), the mistake again arises from confusing three things. In this case, they are *delegation*, *participation* and *aggregation*. They are sometimes looked upon as different names for the same thing, but there are in fact three distinct issues:

— Delegation is, strictly, the delegation of decision-making authority closer to the point of execution.
— Participation, on the other hand, refers to the taking of decisions by groups of people rather than by individuals. They can be taken in many different ways, of course, and there are different degrees of involvement possible (we will be introducing a scale of involvement in a later chapter).
— Aggregation is the principle of dealing, not with individual items, but with logical combinations (aggregates) of them. Thus, setting weekly schedules instead of daily ones is an example of aggregation. So is deal-

ing with the work of a group rather than that of the individuals separately: so is quality testing samples rather than every individual item.

Confusing the three components of the production mode meant automatically assuming, for instance, that assembly lines mean prescribed methods and individual work. Similarly, confusing delegation, participation and aggregation means automatically assuming that centralized systems are necessarily autocratic and concerned with details, while decentralized systems are democratic and simpler.

Moreover, the confusion frequently leads to advantages being claimed for participation that are properly benefits of decentralization, and vice versa. Thus, anticipating an argument to come, *decentralization* by and large speeds up decision-making. As a result, it is occasionally claimed that one of the advantages of *participation* is quicker decisions. And that is a bad mistake.

That the automatically assumed connections are in fact not necessary ones may take a little argument to establish. We can best approach that by looking at each pair of attributes in turn.

Delegation and participation

In a sense, if a manager delegates authority for certain decisions to specific subordinates, he may be said to be increasing the extent to which those subordinates 'participate' in decision-making. But those subordinates themselves may well behave in a highly individualistic manner. They may exercise their new powers without regard for the influence or opinions of their subordinates or their colleagues. Decisions then are not being taken in any more 'participative' a manner than before.

If, on the other hand, the manager does not delegate any specific responsibility or authority, but deliberately sets out to take account of his subordinates' opinions and judgement in forming his decisions, then he may introduce a considerable amount of participation without any delegation. And the effect on both economic and human objectives will be distinctly different in the two cases.

Equally, even a highly participative organization in which, say, all decisions are taken by a simple vote of all concerned may well be highly centralized if that group decision is binding on all members. Again, the results in terms of meeting various objectives will be very different.

In particular, it is worth pointing out that, for the minority, the tyranny of the majority is just as bad as (it may even be worse than) the tyranny of one man. The observation is hardly new: it goes back to Aristotle and probably beyond.

Delegation and participation must therefore be taken as independent principles.

Delegation and aggregation

It is probably generally assumed that if a manager delegates part of his decision-making authority to a number of others (whether groups or individuals), he does so by dealing with aggregates in one form or another. This may often be true, but it is not necessarily so. Even if he delegates authority to deal with them, he may still remain concerned with individual items—wish, for instance, to be kept informed about them if they go wrong, as in 'management by exception'.

And, in any case, in our concern with aggregation we are not looking at what any particular individual does. We are looking at what the system as a whole takes into account. A system is highly aggregative when it deals *only* with total effects and not with component contributions.

If a manager says to a subordinate 'I don't want to know about the details any more. You take care of them,' that of itself does not make the system any more aggregative. The same details are still being looked at, even if by someone else.

Participation and aggregation

Much the same points can be made about the difference between participation and aggregation. It may be less necessary to make them, however. It is easier to see that having groups rather than individuals make decisions says nothing whatsoever about the degree of aggregation of those decisions.

The objectives

Having established that there are three separable dimensions of 'decentralization', delegation, participation and aggregation, we can, as we did in the last chapter, look at their different effects on the human and economic objectives involved.

A summary of their potential impact is presented in Fig. 5.1, which uses a different list of objectives from those in the similar diagram in the last chapter. On the human side, the changes are slight. Product identification has been dropped as being of less importance in the planning context. And self-actualization has been split into two, since job richness and individual influence are differently affected by the different dimensions of decentralization.

On the economic side, however, there is more change. By and large, the objectives listed are derived from the need to strike the balance we mentioned in discussing the resource phase in chapter 3—the balance, that is, between *system power* and *system cost*. System economy we have kept as one objective, but system power is broken down into several components, at two levels.

At the first level, it is broken down into *control* power and *diagnostic*

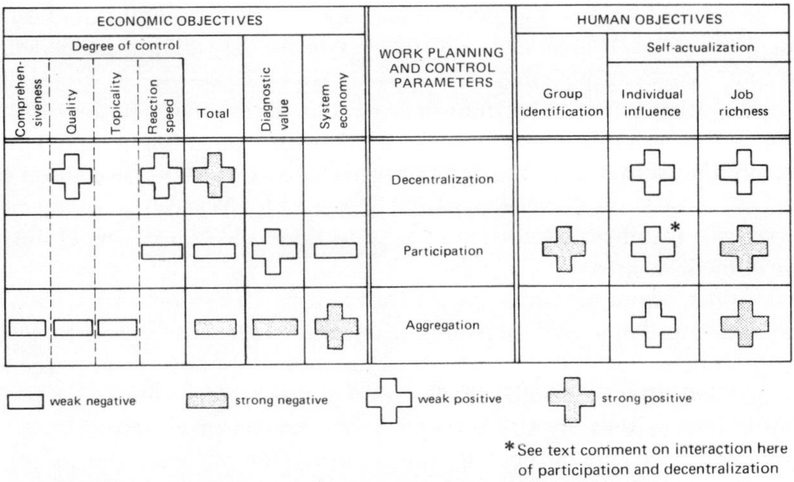

Fig. 5.1. Potential impact of work planning and control parameters.

power. Diagnostic power—the ease and speed with which the system identifies problems and their causes—we take as a single objective. Control power, however, we divide again, this time into four aspects:

— *Comprehensiveness* (the degree to which it takes into account all relevant control factors).
— *Quality* (the accuracy and correctness of the control of the individual factors).
— *Topicality* (the 'up-to-datedness' of the information in the system).
— *Reaction speed* (a measure of the time it takes for the system to respond to a stimulus).

To account for all of these, the left-hand side of Fig. 5.1 unfortunately has to be rather more complicated. Since delegation, participation and aggregation affect each of them differently, they need to be kept separate.

With the objectives explained, we can look in more detail at the impact on them of the three different kinds of decentralization.

On the economic side, increasing *delegation* brings positive advantages to two of the sub-divisions of control power: reaction speed and quality of decision-making. That it increases the speed of decision-making is its main traditional justification. That it also increases the quality of decision-making probably requires an additional comment.

Admittedly, if delegation takes place in a system in which the people involved are not sufficiently competent, the quality of decision-making will drop. But this is simply a matter of control passing from competent to incompetent hands. It would still happen if control passed from a competent centralized authority to an incompetent one. It has nothing to do with the principle of delegation as such. Effectively, we are only talking about delegation to a competent person.

On the other hand, delegation does of itself imply that decisions will be taken by the people best acquainted with the situation at hand. And, by and large, that must in its turn imply an improvement in quality. Delegation can, of course, go too far. 'Over-delegation' can lead to decisions being taken by people with insufficient knowledge of conditions and consequences outside their immediate area.

The other economic factors are unaffected by delegation of itself. What counts in improving or diminishing economy, diagnostic power, topicality or comprehensiveness is the number of decisions that have to be made and the care and attention with which they are taken. The number of people making them (if they are still acting as individuals) is irrelevant.

On the human side of the diagram, delegation in our view has no impact on group identification. The opposite view is sometimes expressed, but it arises from the confusion between delegation and participation. A programme of increasing delegation admittedly increases the number of people making decisions, but only on an individual basis. Any benefit that might be obtained for group identification is, therefore, offset by the natural tendency for each individual decision-maker to view his own affairs as of special importance.

With regard to individual influence and job richness, the benefit to the individual to whom authority is delegated is obvious. However, such individuals are virtually bound to remain in a minority. Therefore, while we have marked the influence on these two objectives as positive, we have described it as minor.

The effects of *participation* are markedly different. On the economic side, it only has a slight influence on the degree of control. This is because, on average, involving several people in each decision must slow down the speed with which they are taken and implemented.

Similarly, the need to involve more than one person in each decision must increase the cost of decision-making. Properly organized, however, this cost need not be too great. We have therefore marked it as a minor influence.

The only positive economic advantage of participation, directly, would appear to be in the field of diagnostic power. Here the involvement of many people ought to speed up the recognition and notification of problems. However, it does not necessarily make it any easier or quicker to find out what is causing the problem, so again the effect is regarded as minor.

On the economic side, therefore, participation has a string of minor effects. On the human side, it produces a major impact on each objective. Group identification is helped, virtually as a matter of definition. Job richness, too, is enhanced by the involvement of the worker in work planning decisions. And security is increased if the worker is involved in some work planning decisions—setting work standards for instance. (Differences between the different kinds of work planning decision is something we will be coming to later.)

With individual influence the situation is a little more complex. Certainly, a participative centralized situation gives the individual more influence than a non-participative one. But at high levels of delegation, the situation is different. A worker with the authority to make decisions for himself has more influence than one simply participating in them with his colleagues.

At any given level of delegation, however, participation increases influence, and markedly so. (It is worth noting here that, in the extreme, complete delegation is incompatible with participation, strictly speaking. Complete delegation means that everyone decides his own affairs for himself. The minimum of participation implies at least two people having to agree on a decision. For practical purposes, however, this can presumably be ignored.'

Aggregation offers mixed blessings on the economic side. Its major benefit is with regard to system economy. Keeping track only of aggregate figures rather than individual items is obviously simpler and cheaper. But there are compensating penalties which affect both diagnostic and control power.

Measuring aggregates alone can slow up, or even prevent the noticing of problems, especially where the aggregation is over time—taking monthly reports instead of weekly ones, for instance. But even other kinds of aggregation—measuring group output instead of individual output, for instance—may also mask incipient problems with one member of the group. Even more strongly, only having aggregate figures makes it more difficult to find the cause of problems when they do show up.

Three sub-divisions of control power are affected. Comprehensiveness is weakened by the use of aggregates. Quality is diminished since the use of aggregates implies approximation instead of accuracy. And where the aggregation is over time, the information used in decision-making tends to be more out-of-date.

Economically, therefore, the cost advantages of aggregation have to be powerful enough to overcome these other disabilities to warrant its introduction. But it also has human benefits. It does not involve interpersonal relationships, so it has no effect on group identification or security but it has a beneficial effect on both job richness and individual influence. If anyone's work is being planned and monitored only on an aggregate basis, it necessarily follows that he has greater freedom in dealing with the individual details. If, for instance, he is being given jobs, and having his performance measured, on a week-to-week basis, he obviously has more freedom than if it is being done on a day-to-day basis.

That concludes our brief assessment of the varying effects of delegation, participation and aggregation on the objectives of the work planning system.

We have also to deal with the question of who participates, or who is given delegated authority. Essentially, there are four parties to the work planning process, each of whom can have varying degrees of authority or right to participate. These are:

- the workers themselves
- their supervisors
- what we will call 'senior management' (the production manager, the works manager, whatever his title may be. For our purposes he is the representative of the central hierarchy. It makes little difference whether he is himself making decisions or passing them on from higher still.)
- specialist staff (the entire work planning process may involve several different specialist departments: preproduction engineering personnel, production control, work study, and so on. It will also include any titular 'planning department'. But for our purposes they may be treated as one group).

It is obvious that it is only the people to whom authority is delegated, or who are given the right to participate, that are beneficially affected. But it is also important to bear in mind that delegation of authority from a participative-minded production manager to a series of autocratic foremen, or to equally autocratic specialist staff is likely to have harmful effects on human objectives. It can easily happen. And it is one more piece of evidence that the claimed advantages of delegation are in no sense automatic.

Unfortunately, in trying to decide how much delegation, participation and aggregation should be introduced into a system, and what roles the four groups mentioned above should play, it is impossible to take the work planning system as a whole. For 'work planning' involves the making of a whole range of different decisions. And the degree of decentralization, of any kind, that is appropriate is unlikely to be the same for each kind of decision. This is because the importance of the various objectives varies from one case to another.

Six major issues

We have already mentioned that work planning decisions can be divided into output-related and input-related classes. To consider the importance of the various objectives, however, we need a finer classification than that. Essentially, there would seem to be six major issues around which all the decisions can be grouped. These are:

(a) *Work standards.* Before any useful work can be done in the planning area, it is essential to know what operations are to take place and how long each of them is to take. It is also necessary to know that output will be of acceptable quality. In other words, if we are to plan at all, we need *standards*. And they need to be of three kinds: standard methods, standard times, and standards of quality. We need to control how things are done, what time they take and how good the results are.

In theory and in practice, there are many ways in which such standards can be set. It is even possible to rely on tradition. To accept, for example, that it will take 15 minutes to do such and such a job, because that is what it has always taken. But for the moment we will leave the merits of different ways of setting standards, while we describe the other planning issues.

(b) *Work goals*. Given an overall plan of campaign (set by the higher-level and more prestigious kinds of planning we mentioned earlier), and given certain set standards, schedules have to be set up. Relative priorities have to be given to different tasks. Production sequences have to be arranged. Given the products and the quantities of each that have to be produced during the next planning period, a programme for producing them has to be created.

This is the process of setting work goals.

(c) *Work administration*. In work administration, the elements involved are the decisions involved in distributing jobs among shops or individual workers, and routing them among different functional areas or departments. (In anything except a fixed-wage system this will also involve making implicit decisions about the payments that workers will get. Fixed wages tend therefore to simplify decision-making in this area, though that is properly a subject for the next chapter.)

(d) *Personnel administration*. The decisions involved under this heading are those affecting the allocation of human resources: the recruitment of new workers and the transfer of excess staff; disciplining workers; training new workers; appointing leaders; rotating and switching workers among tasks, and matching people and jobs according to skill and capability.

(e) *Material and facilities administration*. This includes decisions about ordering materials and tools from store; maintaining buffer stocks; machine maintenance; and quality control of tools and equipment. Effectively, therefore, it parallels the last category, except that it is concerned with material resources, not human ones.

It has also close connections with work goals and work administration except that its time horizons are a little longer: its main concern is that materials and equipment should be available when called for by the programmes.

(f) *Work improvement*. Strictly, this aspect of work planning is different from the previous ones. It is concerned with decisions about improving working methods and conditions, rather than with getting the work out. It is therefore non-routine, and relatively long-term. Nonetheless, we include it here because it is obviously closely connected with categories (a) and (c) above, because it involves essentially the same people, and also because, if neglected or mishandled, the resultant long-term damage can be critically important. Moreover, systems for handling improvements can usually be closely connected with systems for planning routine work.

Some indication of the relative importance of the different objectives on each issue is given in Fig. 5.2. In general, the situation differs too much from case to case to be dogmatic, but some things are nearly always true. Thus, the need for speed is lower in the setting of work standards (which by definition happens only periodically) than it is in work or materials administration, which are ongoing tasks.

Similarly, the need to provide qualified staff varies from issue to issue, which will make a considerable difference in answering the question: 'Who should participate?' For instance, the need for specialist staff in calculating work standards is normally higher than it is in personnel administration. Workers generally, and their supervisors, are likely to be just as competent to evaluate the capabilities of their colleagues as specialists would be.

How to simplify: MOST, MTM and SMAC

As we pointed out before, work standards are at the heart of the entire work planning process. Unless they are competently set, the system is bound to be inefficient. That requires a degree of professional competence in the people who set them: though in controlling against them, there are considerable benefits to be gained from participation, especially in the area of quality, as the television case in the last chapter illustrated.

Unless work standards are reckoned to be objective and fair, the system based on them is bound to be resented by the people involved in it. Then it will again be inefficient. Necessary too, therefore, is a degree of collaboration in standard-setting by the people involved, who are unlikely to have the professional competence demanded by the last paragraph. It also follows that the methods used to set standards need to be reasonably simple to understand.

The development of new ways of setting them has to be heavily biased towards simplicity (*through* sophistication, not at the expense of it) and speed in operation—without forsaking accuracy. An example of this trend is the Maynard Operation Sequence Technique (MOST). MOST is developed from earlier Methods Time Measurement (MTM) techniques, but it is, nevertheless, an independent technique which can be put to use after only a couple of training days.

The benefits of the technique in speed and simplicity are evident from the

Economic objectives							Human objectives		
	Degree of control					Work planning issue		Self-actualization	
Compre-hensiveness	Quality	Topical-ity	Reaction speed	Diagnostic value	System economy		Group identification	Individual influence	Job richness
mod.	high	low	low	high	mod.	work standards	low	high	mod.
high	mod.	high	high	high	mod.	work goals	high	high	high
high	mod.	high	high	high	high	work admin.	high	high	mod.
high	mod.	high	mod.	high	high	personnel admin.	high	high	mod.
high	mod.	high	high	low	high	materials admin.	low	low	mod.
mod.	high	low	low	low	mod.	work improvement	mod.	high	high

Note: 'high', 'moderate' and 'low' should be read comparatively. 'Low' does not mean unimportant.

Fig. 5.2. Relative importance of different objectives for different planning issues.

following table, which compares MOST with three different levels of analysis using more traditional systems.

	To define and measure one hour's work	
	Documentation (pages)	*Engineering time (hours)*
MTM-1	59	300–400
MTM-2	41	100
MTM-3	28	30
MOST	3	5–10

At the same time, in the majority of industries, MOST represents no loss of accuracy compared to MTM-2 and a substantial improvement in accuracy compared to MTM-3. And the considerable reduction in learning time makes it especially suitable for participative schemes.

While a technique like MOST simplifies one issue of the planning system, there are techniques that can be used to simplify the whole system. The benefits that group technology can confer on the production mode designer we have already discussed. Properly handled, and added to other techniques for simplification and problem-solving, as in the Maynard Selective Manufacturing and Control programme (SMAC), group technology can also make improvements to the manageability of batch production operations, simplifying what can easily become an intractable planning and control problem.

The techniques of group technology are essentially taxonomic. That is, they are used to categorize products and their component parts in order to minimize their variety, and in order to impose common manufacturing requirements in terms of tools and fixtures. It can thus produce considerable benefits in direct ways, through savings in materials, set-up times, operation times, preproduction engineering and the like—all things not immediately part of work planning. It does so especially when coupled, as in a SMAC programme, with what is sometimes known as ABC-analysis and sometimes as the application of 'Pareto's Law' (that 80% of the input goes on producing 20% of the output, though that is not how Pareto phrased it). In this way, it leads to an improved production mode, both from the economic and from the human point of view.

But it also produces a major overall simplification in manageability, as sketchily indicated in Fig. 5.3. Essentially, it reduces the number of elements that have to be planned for. But it also, importantly, replaces a series of *ad hoc* demands for varying, unrelated operations with repeating demands for similar work. The runs themselves may be no longer, but they repeat. So, therefore, do the decisions involved, lessening the amount of information and the degree of expertise needed to make them.

Even more powerfully, a critical analysis of the economic importance of

	Selective Manufacturing And Control will have this impact	
Materials planning and control	Reduced throughput times. Less work-in-progress. Reliable delivery by due date. Decreased production areas.	Machines close together. Complete process under one supervisor. Short ordering cycle. Less inter-process delay. Less buffer inventory.
Direct and indirect labour	Reduced set-up times. Reduced operation times. Reduced transportation and handling times.	Process-oriented production reduces set-up and permits high-volume production methods and tooling. Repetitive routing, machines placed in process sequence.
Production planning and control	Simpler, decentralized planning. Reduced order handling cost.	Fewer load centres. Fewer planning and control transactions. Fewer active orders.
Work organization	Improved worker motivation. Higher resource utilization. Better quality.	Permits delegation of task distribution, worker reassignment and quality control. One foreman controls complete process.
Preproduction engineering	Improved motivation for preproduction engineering personnel.	Eliminates largely the routine preproduction engineering. Preproduction engineering by complete process rather than by single operation.
	Lower preproduction engineering cost.	Eliminates routing. Reduces need for methods specification and standard-setting. Selective engineering based on cost/benefit: Low volume/value = simplified engineering; high volume/value = thorough engineering.

Fig. 5.3.

parts will often lead to the adoption of a simplified specification, control system, and production regime for parts of low value, with substantial unit cost reductions in the preproduction departments. For some such parts, the requirements for specification and control can be delegated to a purpose-designed section of the shop with selected supervision, so that the need for specialist planning diminishes to vanishing point. Meanwhile, it becomes more feasible to use intensified production engineering on the remainder of the range—that part of it that has the higher value.

Planning and the computer

One approach to the planning problem in the modern production system has been to attempt in one way or another to simplify it. The alternative approach has been to accept the complexity, and computerize it.

In view of the complexity of their problems, it is not surprising that so many planning system designers greeted the advent of the computer with enthusiasm. Nor is it surprising that, throughout the 'sixties, huge efforts were made to cope with the problem through the development of ever more sophisticated systems.

Unfortunately, with few exceptions, the success of computerized work planning systems has not matched their promise. Multitudinous reasons have been advanced. Four basic ones would however seem to cover most of the ground. They are:

(a) Computerization of what exists

(By 'what exists' we mean either the planning mode that exists, or the production mode that exists.)

This is the reason most often put forward by computer people themselves. Basically, it means the charge that systems have not been designed around the computer, and therefore do not make maximum use of its capabilities. Instead, the machine has merely been used to try and speed up and make more reliable the existing system. Or, as one enlightened EDP manager put it: 'There's just no point fitting wheels under a sleigh.'

(b) The incompetence of the computer

The charge here is one that is, perhaps surprisingly, levelled as often by computer specialists as by laymen. It is not the accusation that the analysts and programmers involved have been incompetent, but the assertion that the computer itself is not as powerful as it has been made out to be—that, in effect, computers have been oversold, at least in view of their present capabilities, that expectations have been artificially high, and that therefore tasks have been attempted that are not as yet achievable.

There is more in this charge than is perhaps generally realized. The information-processing capacity of modern computers is admittedly high. But in companies with complex product and demand structures and production modes, information-processing requirements may well be higher.

This is especially true in situations where much of the information involved is of a kind that is difficult or expensive to store in digital form, as it often is where drawings are important. And it is accentuated particularly in complex systems where a change in any one factor may (or may not) affect any or all of the others. In fact, no administrative application is more likely to over-extend the computer than work planning.

Both charges (a) and (b) have in common that they imply that someone has made a technical mistake. They are therefore susceptible to technical solution, which makes them of less relevance to our general concern in this book. The remaining two charges, however, bring us back to the theme of changing attitudes and aspirations on the part of the workforce, and to the mingling of human and economic objectives generally.

(c) Lack of understanding

This is a charge that falls neatly into two halves. On the one hand, the assertion is that the people required to work the system lack sufficient understanding of computers. On the other, it is that the computer specialists lack sufficient understanding of the real-life situation, including the people involved. Both ways, however, it can be viewed as a problem in *interpretation*. That is, it is a problem of reconciling the kind and amount of information that people habitually use, with the kind and amount required by the computer.

Typically, the computer requires complete and rigidly determined inputs if it is to operate properly. This seemingly simple and straightforward prerequisite, however, is a towering hurdle in practice. The human being is used to accepting (and reacting intelligently to) incomplete or 'obviously' wrong information. It is only natural that he should expect the computer to show the same tolerance and understanding.

Human beings note patterns, and disruptions to patterns, quickly and easily. But they are inadequate at handling large quantities of detail. Computers thrive on the detail, but they have difficulty in recognizing patterns—especially unforeseen patterns.

Perhaps most concisely, the problem is the computer's failure to recognize the 'obvious'. Everybody probably has their own pet anecdote illustrating the potential humour in such a situation. Our own favourite concerns a refrigerator manufacturer with a computerized stock control and ordering system. The manufacturer, so the story goes, brought in a range of pink refrigerators, which did not sell. Instead, they mounted up in the warehouse until they reached the programmed stock level. After impressive but vain efforts, the sales department gave up on the range, and the entire stock was sold off cheap as a job lot.

Promptly the next week, the warehouse was stocked up with pink refrigerators again. . . .

This is different from the previous two charges. It is not simply a matter of someone's incompetence. Instead, it brings us close to our central concern. In the first place, it must be an essential principle that the effectiveness of any planning system, computerized or not, is limited by the degree to which the people operating it (and controlled by it) understand its mechanisms and purpose. Whatever the situation may have been at one time, today unquestioning acceptance of instructions can no longer be taken for granted. The extent to which people carry out a plan is therefore dependent on the extent to which they understand it.

The result otherwise may well be that two planning systems come into use. The official one, designed by the designer, is carried out mechanically by people going through the motions. But the unofficial one (probably the old one) is the one that counts. It is the one they carry out because they rely on it—because they understand it.

(d) Over-centralization

This fourth and last charge brings us to the heart of the problem of the changing role of the worker. To make the point properly, it is worth going back a little in time to the period immediately before the advent of the computer.

For some while, traditionally centralized (in all three of the senses: little delegation, participation or aggregation) planning and control systems had been under pressure for practical as well as psychological reasons. Largely because of the search for economies of scale and the marketing benefits of product differentiation, production operations had been steadily growing in size and complexity. And centralized systems had shown up as rather unwieldy things.

They slowed up decision-making, because of the increasing number of hierarchical levels between the decision-maker and the action point as the organization grew larger. They also complicated considerably the speed of information flow even between units on the same level of the hierarchy.

The result was a drive for decentralization, especially in the sense of increasing delegation. Delegating authority down the ladder was seen as the key to greater flexibility (especially in coping with unforeseen changes) and decreasing lead time. Consequently, it also led to improvements in such areas as machine utilization, work-in-progress levels, and so on.

The first impact of the computer, on the surface, was to remove this obstacle to centralization. The computer-aided centralized planning and control system was assumed to be even faster in action, and better able to cope with detail than systems with a high degree of delegation. The computer therefore promised a new lease of life to the autocratic manager just as his position was being undermined. And it was undoubtedly frequently used for that purpose.

But for this reason it ran full tilt up against the changes in workers' attitudes to their work that we have already been concerned with. It failed to take into account the growing demand for such things as more satisfying jobs and greater influence in decision-making. It made things worse in some respects. One could always argue with even an autocratic superior, and thus maintain some self-respect. One cannot argue with a computer.

No-one could claim that this is a new observation. There has already been much reaction against the idea of a computer simply as a prop to centralization. But it does emphasize our general point: the need for the planning system designer, computer-aided or not, to make his system serve both economic and human ends.

Of the four charges against the computer, therefore, two are technical and two organizational. Computerizing what exists and overestimating the power of the computer are failings that can be avoided by proper professional expertise. The other two objections, however, are fundamentally organizational. Lack of understanding (on both sides), while it may look like a problem in technical instruction merely, can only be effectively coped with by an extension of participation by allowing specialists and non-specialists to learn about each other through working together. And the dangers of over-centralization are only avoidable through maintaining a reasonable degree of delegation—something that has been facilitated by technical developments in computer hardware—for instance, 'intelligent terminals'.

The only kind of decentralization that the computer properly diminishes the need for is the third: aggregation. It is generally true that computer-aided systems can afford to be less aggregative. With the computer it becomes relatively cheaper to hold more detail in the system. The mistake is to assume that that justifies cutting down on delegation and participation.

What has tended to happen in practice, and how companies have moved towards solving their computer problems, is reasonably summarized by the experience of one Dutch company. People who have lived through it divide the history of that company's experience with computers into four stages.

The first was the era when 'the saints came marching in'. This was the period of wild excitement, glowing forecasts, and a sense of the millennium. Unfortunately, it was also a period of drastic underestimation of the time and effort that was going to be necessary—not only to design the systems but to compile sufficiently complete and reliable basic data on product and production structures.

What was showing up was the set of problems we have described under charges (a) and (b): computerization of what exists and overestimation of the computer power. The reaction was to move into the second stage. That stage was marked by strenuous attempts to hire and apply top technical talent, seeking to solve what was still seen as a technical problem by the deployment of professional expertise.

The result was that they ran into charge (c) problems: lack of understanding. The recipe for solving those that evolved in the third stage was, correctly enough, to increase participation. Project groups were set up for each computerization project to include both technicians and operations people. The trouble then became that nearly all the projects took too long to implement. As we have already pointed out, the effect of participation generally is to slow things down.

Stage four evolved in an attempt to speed things up to a practical time scale. And it involved increasing the degree of delegation. In stage four, the present stage, the purpose and scope of computerization projects are entirely defined by the users. And no project is approvable if it is scheduled to take longer than six months to implement.

The result is that many of the early dreams have faded, and most of the grandiose schemes been abandoned. But, instead, a reasonably human approach has for the first time in that company produced an economically effective way of utilizing the computer.

That disquisition about the impact of computers ends our argument for this chapter. To illustrate the case we have made, we can turn to some brief examples.

Example 1. Building materials

A rather unusual situation is provided by a building materials company in Sweden, using a high degree of automation to excavate and process clay. Inherent in the conditions is a much higher degree of fluctuation in various key factors than is normally true of automated process industries.

Production, being from open pits, is very dependent on the weather. Capacity in difficult winter conditions is only about half of that in the best summer conditions. Demand also varies considerably. Over the year it varies with the change in seasons and the annual building holidays. Over the week it also varies, with most shipments taking place on Thursdays and Fridays. Even during the day, shipments tend to be greater in the evening than in the daytime.

Restoration work in the pits is also seasonal, being more appropriate to the summer months. The capacity of the clay processing unit also varies with the water content of the clay.

Finally, there are the sources of variation common to most process industries. The need for repair and maintenance work fluctuates considerably. Since the number of men working on the automated line is small, variations due to absence, sickness, training or holidays also gain increased importance.

Studies indicated that, to cope with the average work volume, 150 man-shifts per week would be required. The work content of each of those shifts would, however, vary considerably and unpredictably. For them to be carried out by 30 men working a five-shift week would require a high degree of variation in the tasks each man would have to do.

It was therefore decided to set up six shift-teams of five men each, but to ensure that between them the men in each team would command all the skills necessary for all production and maintenance tasks. Each team was provided with at least two men trained in each task.

On any weekday, two teams would work the morning shift, two the afternoon and one would work at night. The sixth would be off. On Saturdays and Sundays, only one team would work at a time.

As a result, there would never be more than two five-man teams working at any one time. In those circumstances it was decided that all supervision, apart from the presence of the plant manager himself, could be dispensed with.

In terms of the framework described in this chapter, the resulting situation can be analysed as:

Work standards have not in fact been of central importance. Where questions affecting them arise, they are dealt with at regular weekly planning meetings, chaired by the local manager, and including a representative of each team. The job of representative is intended to rotate within the teams. On this issue, therefore, participation is high, though, since all the teams are bound by the meetings decisions, delegation is low.

Work goals for any particular shift are highly dependent on conditions at the start of each shift. Responsibility for setting the goals for each shift can therefore be left with the group performing it. But it is also essential for the incoming group to gather from the outgoing one the information it needs to decide what it is going to do on the shift. This has been achieved through overlapping the shift relief times.

Work, *personnel*, and *materials administration*, at least for short-term matters, are dealt with on the same basis. Responsibility is given to the group, but continuity of information is ensured by overlapping relief times.

On these last issues therefore the emphasis is on delegation *to* the groups and *participation* within them. Inter-group participation also facilitates the flow of information.

Finally, *work improvement* issues are dealt with on the same basis as work standards at the weekly planning meetings. These also deal with longer-term personnel administration matters, and the provision of information about the company's general activities. Such matters are, strictly speaking, not part of work planning, but are more the kind of thing chapter 8 will be dealing with. However, it is worth noting that devices used to solve work planning problems may break fresh ground in such a way that they help with other problems too.

To complete the picture, a three-man team was also set up to aid the local manager. This includes two specialist technicians (one for process techniques, one for maintenance matters) who deal with problems outside the competence of members of the production groups. They are also responsible for staff training, and deputizing for the manager when necessary. The third member of the team is a clerk who carries out normal office duties and also is involved in making shipments and storage depot work.

Example 2. Customer service

In one form or another, the following example is a familiar one. The specific facts refer to a company working in the installation, repair and maintenance of heating and sanitary equipment. But its basic situation is a common one in service organizations generally.

Before the reorganization, the company's 18 fitters gathered each morning to receive orders from the foreman. He, in turn, was provided with them by the order clerk. Assignments were allocated by the foremen to specific fitters, who then collected a bag with the hand tools assumed to be necessary, and took from stores any additional equipment or material required. The fitter then went off to the address given on the order note. And, when things were working properly, he performed the necessary work.

Unfortunately, many things could go wrong. The house or flat, or the necessary valves, might turn out to be inaccessible. The tools and spares taken along might turn out to be unsuitable. Repeated calls would then be necessary, with consequent complications to the planning system. The complications resulted both in increased costs, and ever lengthening work queues. There was as a result a steady deterioration in customer relations. To salvage these, reorganization was planned.

Under the reorganization, the daily work allocation routine was abandoned. Instead, the fitters were each supplied with a van equipped with two-way radio. Each van also carried a complete range of tools and materials.

On receipt of a customer order, it is now possible to contact a fitter immediately. If there is no one directly available, the job can be allocated to the fitter able to undertake it soonest. Via radio and telephone, it is also possible to arrange a suitable time with the customer and fitter, and to be reasonably confident that the van that arrives will be suitably equipped.

Effectively, therefore, for most of the work planning issues, the new system relies on almost complete delegation. One might almost say that a service department with 18 operatives has been replaced by 18 service departments.

Participation is minimal. To allow for exchange of information, however, and also for the making of decisions (on work improvement and standards, for instance) that affect everybody, weekly 'production meetings' are held. These are attended by all the fitters and the foreman. As in the previous example, such meetings have become the standard forum for discussing long-term issues outside work planning, of the sort we will be dealing with later.

6. Structuring pay systems

Attitudes to pay schemes. Down with piecework? Sweden's long, hot winter. The job, the man, the time, the output. And profit-sharing too. Which does what? Mixing the blend.

Conventionally, pay systems have only been of limited concern to production system designers. Generally, pay systems are designed and introduced taking the production system as given; and vice versa—changes to the production system usually take place against the background of an established pay system. Only when the two change together can they be linked. And even then, the role of the production system designer is likely to be limited to working out standards for piecework and other output-related payments.

Giving the subject a major place here, therefore, requires some justification. It is not hard to find, however.

Central to it is the fact that it is not enough for any designer to produce a system that works on paper. It also has to work when people are operating it. We have already argued that that is why the designer has to take human objectives into account. And it would be pointless to do that without considering material rewards.

For, whatever emphasis is placed on the intangible reasons for working, the earning of a living remains a central one. And there is more to it. Granted it may be true that, ultimately, any firm is likely to find itself paying average rates for the particular locality in which it finds its labour. But what it gets for those average rates can vary considerably.

Poorly constructed or poorly administered systems can lead to a waste of time and effort, as well as having a bad effect on human relations. On the other hand, well constructed pay schemes are, in effect, management systems. They not only stimulate performance but they also increase workers' and supervisors' direct interest in keeping the whole production system going. They provide a motive for keeping service and maintenance functions alert and effective, for spotting faults and bringing them to notice, and, in general, for helping to control the system more effectively and efficiently.

For the planning and control system designer, a good pay scheme provides a built-in early warning system.

Moreover, the data requirements of planning and control overlap those of payment systems. Unless they are designed as a whole, the result is likely to be confusion, if not conflict. Ideally, the data for the payment system should flow automatically from that required for production management anyway. In fact, even where incentive schemes are ruled out for some reason, the data on which they would need to be based may still have to be collected.

It also means that to install an incentive scheme requires a considerable amount of methods engineering and work measurement for the determination of standards. It is ironic that the costs of that study and the consequent wage administration are often seen as a disadvantage. In fact, the carrying out of such work should be one of the major benefits of installing incentives.

We will be returning to such topics in more detail later. For the moment, having established a case for our concern with the subject, we can look at what has been the dominant movement in the field, at least in Western Europe, over the last several years.

Attitudes to pay schemes

Sweden is the traditional haven of industrial peace. But in 1969–70 it went through what Rolf Lindholm of the Swedish Employers' Confederation (SAF) calls its 'long, hot winter' of discontent. A series of wild-cat strikes spread through the economy.

The transmission rate of the unrest was faster than a biological plague. One weekend, one of the country's major employers settled a strike by conceding a major wage increase of 1·4 Swedish crowns an hour. The settlement was announced on the radio at 6.15 on the Monday morning. By 7.15, the Employers' Confederation had been notified of 10 more strikes. By the end of the day there were stoppages in over 20 other companies. In each case, first among the employees' demands was a pay increase of (roughly speaking) 1·4 crowns an hour.

More interestingly from the point of view of this chapter, however, in second place on each list was a demand for the ending of incentive schemes and their replacement by fixed-wage systems.

It ought to be remembered that, up to that time, Sweden had been one of the great bastions of the pure piecework system. For a long time, the percentage of manual workers being paid by performance was higher in Sweden than anywhere in the world. In the early 'sixties, Sweden was overhauled by the USSR and some other Eastern European countries, but at the end of the decade more Swedish workers were paid by results than anywhere in the non-Communist world.

(Interestingly, perhaps, in the Chinese part of the Communist world, the situation is very different. Russia is the world's biggest user of incentive payment, China far and away the lowest.)

The Swedish workers who demanded the abandonment of incentives were therefore challenging something very firmly entrenched in their society. But in doing so they were very much in line with contemporary trends. In most countries in the West (except perhaps the US and Japan, which we will return to in a moment) the institution of piecework in particular and payment-by-results in general was coming under fire.

It was not always the workers who demanded its abolition. In the United Kingdom, in particular, the pressure to end incentives usually originated with management. Indeed, it was frequently resisted by the unions, on the ground that it weakened their bargaining power. Sometimes, as ICI found in their original attempts to introduce measured day work in 1966–7, something that had been agreed with the unions could well be rejected at the shopfloor.

In Japan, where there is almost total dependence on paying people strictly according to rank and seniority, the question could hardly arise. In the US, too, the campaign never seems to have become a serious issue. The Conference Board of New York, in its regular assessment of the industrial scene for subjects worth reporting on, has never felt the subject warranted study. Requests to Maynards for assistance in developing pay systems are about evenly split between those wanting to introduce incentives and those wanting to eliminate them.

US companies in general appear more or less content with what they already have (whichever it happens to be). What changes there have been have largely been confined to the improvement or replacement of systems that for local and technical reasons were not working too well. At least for several decades, there has been no ideological trend corresponding to that in Europe.

(It may be important here that the variety of payment systems in use in the US is probably wider than anywhere. The variation between industries and between geographical areas is also high. Overall, the US rates fairly low as a user of payment-by-results. But this masks the fact that in some areas, like Pennsylvania, the proportion is extremely high, whereas in, say, California it is extremely low.)

Transnational generalizations are therefore difficult to make. But there is one that holds good almost everywhere. That is that the scientific management school was, by and large, *for* payment by results, and the behavioural science school was, on the whole, *against* it. (Union attitudes were therefore very much coloured by their attitude to behavioural science: by whether they saw it as something that could only benefit the worker, or as something that merely cloaked an extension of management exploitation.)

Down with piecework?

The behavioural science argument usually went like this. Payment by results automatically implies a fluctuating level of earnings. That level is dependent not only on the worker's efforts, but also on other circumstances: materials' shortages, machine breakdowns, the pace of other workers, and so on. Such fluctuating incomes mean insecurity.

In addition, the worker on payment by results, effectively, is not being trusted. Having to justify each week's wages with his output, he may feel belittled or discriminated against in comparison with the white-collar worker on his fixed weekly salary. And that either creates resentment in him or worsens any resentment he already has.

Insecurity and resentment, the argument runs, both breed alienation. In the long run that leads to poor productivity and a demotivated workforce.

There were other arguments against incentives too: perhaps more tangible ones. It did not have to be that way, but traditionally payment by results meant payment for quantity alone. Other relevant 'results' over which workers may have had some measure of control were rarely considered: material costs, repair and maintenance costs, quality costs and many others. In some industries, the impact on overall production economics by these 'overlooked' factors was disturbing. In such industries, many managers felt that concern for those factors was being positively harmed by overstimulating the workers' interest in quantity.

In other industries, the work content of blue-collar jobs was becoming much more a matter of the control of equipment, making output a much less direct function of effort.

In many situations, too, the proportion of blue-collar workers was dropping. In some, white-collar workers were becoming the majority. And even among blue-collar, the proportion of maintenance and other indirect workers was steadily increasing.

In general, therefore, the number of workers for whom traditional incentive schemes had been designed was diminishing. The usefulness of the schemes was becoming steadily smaller. And the task of designing them to fit the new-style workforce was becoming steadily more complex.

A different line of argument evolved from the work of some researchers like Professor Lupton of Manchester. He provided evidence to confirm what some line managers already believed: that incentive payments had no effect one way or the other. After working for some time on the shopfloor himself, Lupton brought attention to the importance of what he called *systematic soldiering*: deliberate manipulation of the output level to ensure, not maximum earnings, but stability. Informally, and perhaps without any deliberate collusion, work groups will establish what seems to them a satisfactory wage level. They then match their output to what they want to earn. For wage increases, they will then rely on union bargaining over the rate, rather than on attempts to produce more.

Lupton called this 'soldiering' because of the parallel with reported experiences in the First World War. Then, front line troops in static trench situations would tacitly establish activity levels high enough to satisfy their officers, but not so high as to threaten serious damage to each other. For instance, regular schedules would be set up, and the same targets brought under attack by artillery or mortars at the same time each day. Thus the appearance of action was produced while the other side was allowed to make sure it was under cover at the appropriate times.

This line of argument is typical of many unjustified attacks on the principle of incentive pay schemes. What is really being attacked—and rightfully—is technical deficiency in specific pay schemes. The 'soldiering' effect could not exist within a sound scheme based on correct standards.

Finally, there were criticisms that even proponents of payment by results had to admit were justified. These were that such schemes tended to be rigid, and to hold back natural developments, and were frequently overcomplicated and difficult to administer. They could be expensive to set up, and create resentments through errors in calculation or differences in interpretation. Sometimes they led to corruption. And, while the standards they were originally based on may have been impeccable, in the course of time they tended to deteriorate unless carefully maintained.

Payments based on them could easily, therefore, degenerate into runaway wage costs, and lead to a virtually continuous state of renegotiation and consequently difficult labour relations. A powerful argument in the United Kingdom particularly, this was the major reason behind British Leyland's long and tortuous attempts to eliminate piecework.

Thus, there were plenty of reasons for both management and workers to want to see the back of such systems. Only a few voices could be heard still arguing the other way. Some unions did not want to give up favourable 'incentive schemes' that happened to provide a comfortable framework for 'soldiering'. Some diehard managements saw no reliable substitute for incentives in maintaining their laboriously achieved productivity records.

Some offered more subtle arguments, such as that people like to be able to control their own earnings—like to be able to earn a little more when they want it, and relax when they can afford it. Still others, perhaps more realistically, maintained that if you were going to have to pay people more anyway, it was better to do it in return for more work. But they all suddenly sounded very oldfashioned.

Sweden's experience

So it is interesting to look at the practical results of Sweden's 'long, hot winter'. The SAF subsequently collected data from 36 companies that had made the switch from piecework to fixed wage systems. The results, summarized in Fig. 6.1 under the heading 'A', were remarkably consistent.

Only two of the firms showed productivity increases (in volume per working hour), and then no more than would have been expected from previous trends. The rest reported drops in productivity ranging from very slight to over 25%. Fifteen of the companies reported drops of between 10 and 20%.

Other statistics were no more promising. Thirty firms reported unchanged quality, 32 unchanged labour turnover, and 33 unchanged absenteeism. Five firms reported improved quality against one worse. Two reported better turnover and two worse. And one reported a better absenteeism record, while two were worse.

No matter whether the variations were for the better or for the worse, they were reported by the companies as being due to the change in the wage system. Some fairly subtle rationalizations were put forward. One company suggested that absenteeism increased after the introduction of fixed wages because workers felt less committed to their fellow workers. Under piecework, if a worker took a day off, his fellow-workers' earnings suffered. Under fixed wages, only he suffered from being absent. He therefore felt freer to take the time off.

All the companies, however, reported a calmer working climate. There was less tension and fewer disputes. That part of the argument against piecework looked impressive. But the consequent argument that the calmer climate would result in increased productivity was made to look like wishful thinking.

The fact of the matter appeared to be that people will not work as hard when their pay is not related to their output. And that, roughly speaking, from the turnover record, just about as many people prefer fixed wages as prefer output-related ones, with most people not caring very much either way.

These impressions were heightened by two studies the SAF made at the same time, and which are also summarized in Fig. 6.1. These were of the effect of introducing 'premium' systems.

Premium systems were defined by the SAF as systems in which the principal part (70 to 90%) of the individual's wage is paid as a fixed item. Then an incentive increment of 10 to 30% of the total depends on the productivity of the individual or the group. (In contrast, 'piecework' systems were defined as ones in which at least 50% of the wage was variable.)

Of the companies studied, 21 had changed from piecework to premium systems, and 16 had changed from fixed wages. All of them reported improvements in productivity (Fig. 6.1, sections B and C).

Among those that changed from piecework, average improvements were of the order of 10 to 15%. Among those that changed from fixed wages the average was over 30%. Four of the companies in this last group were also

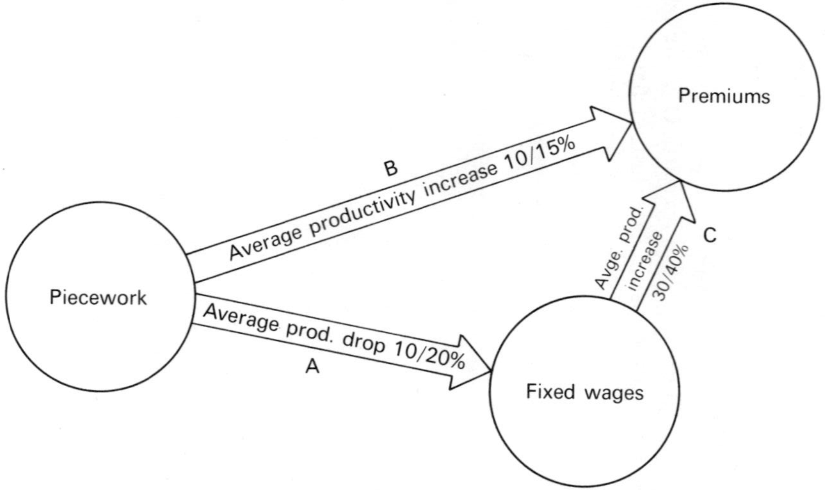

Type of change	Number of companies reporting																	
	Productive efficiency								Quality			Absenteeism			Personnel turnover			
	Down				Up													
	Over 30%	20/30	10/20	0/10	0/10	10/20	20/30	Over 30%	Worse	Same	Better	Worse	Same	Better	Worse	Same	Better	
A	2	7	15	10	2				1	30	5	2	33	1	2	32	2	
B					10	6	5			18	3		20	1		19	2	
C					2	2	5	7	1	14	1		16			15	1	

Fig. 6.1. Effects of changing pay systems. Study of 73 Swedish firms by Swedish Employers' Confederation.

among those that had changed from piecework to fixed wages at an earlier stage. Their gains from the switch to premiums more than recouped their earlier losses.

On the question of quality there was again little change in both groups. The same was true of both absenteeism and turnover, though, as far as these two factors were concerned, no company reported a worsened situation.

There is a difficulty in interpreting the results, of course. Some of the reported improvements may have been due to a 'Hawthorne' effect—people working harder because management was taking an interest in them. Or there may have been other changes that coincided with the effect of the wage changes. But, nevertheless, taking the studies together indicates that productivity is increased when earnings are output-related, but that the process can go too far. Diminishing returns set in somewhere around the point where the variable part of income is around 30% of the total.

The studies also indicate that, in general, the form of the payment system makes little difference to the way that workers feel about their jobs—at least not enough to make them leave, or their attention to quality drop.

While these assertions are based directly on Swedish experience, they are not ones that it would be easy to quarrel with from experience in other countries. There may be variations in emphasis from one country to another, but the overall picture remains fairly consistent.

A recent survey by the American Institute of Industrial Engineers of 453 US firms produced different figures in detail, but the same picture in general. Firms changing from no measurement to a system of measured day work reported an increase in productivity of 24·7%, while those changing from no measurement to incentives reported an increase of 51·5%. These figures, however, were possibly biased in favour of measured day work by the fact that those firms that introduced it (approximately half the sample) had a lower level of productivity before the change than those that introduced incentives.

After allowing for that difference (by reckoning the improvement from the pre-change average productivity of the whole sample), the Institute revised the figures to report an improvement of 14·6% on the introduction of measured day work and an improvement of 63·8% on the introduction of incentives.

There were no data on changing from measured day work to incentives, or vice versa.

It may seem that what is needed is a proper study of the whole matter. It would indeed be an advantage if we had the results of a fully systematic and empirical attempt to relate the way workers feel about—and perform—their jobs to the way they are paid. It has been tried, by one employers' federation that prefers not to be named. But, unfortunately, in the end, their elaborate and expensive attempt had to be abandoned, after it had bogged down in a morass of multitudinous interacting variables.

However, without going to those lengths, it is possible to advance a thesis which is in line with those we have advanced in previous chapters. It takes account of both the similarity between countries and the variations, and provides a more useful basis for practical development than the simple statements above.

This is that there are several basic human goals that the payment system can help to satisfy, and that there are several different components in the income package, each of which has a different effect on each goal.

One would therefore expect responses to the same payment systems to be similar everywhere. But the relative strength of those different goals varies from individual to individual and from community to community. One must also expect, therefore, detailed variations in those generally similar responses.

As in previous chapters, we need to identify those several components of the system and the objectives the system is intended to serve. We can then relate one set to the other.

Key components and objectives

The payment that any worker receives at the end of the week or the month is determined by some combination of four factors. In any particular case, some of the factors may not be relevant, but the potential determinants are: the job that he does; who he is; the output he has achieved; and the time he has spent working.

In addition, there is the possibility that he will occasionally receive bonuses that are in some way dependent on the financial success of the company. These may vary from the paternalistic and arbitrary 'Christmas bonuses' still given by some firms, to formal and settled profit-sharing or productivity gain-sharing agreements or even equity participation. What they all have in common, and what separates them from the other four factors, is that they are not dependent on the worker's own contribution to the success of the company.

We will therefore have five components (the job, the man, the output, and the time spent working, plus bonuses) which need to be related to the various possible objectives. A lot more needs to be said about them. But before we do so it is worth looking at what objectives the payment system can help to serve.

On the economic side, it should be pointed out first that payment-by-results schemes can be linked to whatever results one pleases. In a sense, therefore, any economic objective can be affected by the payment scheme. All one needs to do is include it in the formula that measures results. Incentives can be tied to economic use of materials or time, or to quality, just as much as to quantity (though not always so easily). But this is not the kind of effect we want to look at. What we are concerned with here are the overall effects that the payments sub-system may have on the general economic objectives of the whole production system.

The major ways in which it does that are through its effect on the system's

— *Flexibility* through the ease with which it allows people's jobs to be changed.
— *Self-correction* through the way in which it stimulates (or fails to stimulate) workers and, indirectly, supervisors to take corrective action themselves.
— *Stability* through the way in which it increases or decreases tensions that may lead to disruption.
— *Performance* through its direct action to stimulate workers' efforts.

Of the three factors, utilization, methods, and performance, which we introduced in chapter 3, performance is therefore explicitly mentioned. The impact of flexibility is mainly on utilization, though it also affects methods. And self-correction and stability make their major impact on utilization.

On the human side, we can also break the list of objectives down into four. *Security*, in the financial sense, is an obvious one. *Group identification* can also be kept with no qualification. Self-actualization, however, we need to break down into two, which are differently affected by the different possible components of the wage packet. These are *individual influence*, the power the individual has to affect his short-term earnings; and *self-development*, the power the system gives him to bring about long-term improvements in his circumstances.

Figure 6.2 indicates roughly how each of the components of earnings are likely to affect each objective. Of the five components, two are classifiable as *fixed* (pay for the job and pay for the man). Three are classified as *variable* (pay for time spent working and for output, and bonuses). Within each group, the effects are, not surprisingly, somewhat similar. In describing the impact made by each component, we can therefore simplify things a little by taking group effects together.

Fixed components

The attribute that these two components have in common is that earnings are fixed on a relatively long-term basis. They only change when the man moves from one job to another, when there is some change in his personal circumstances, or, of course, when negotiation results in a new rate being set.

Week-to-week and month-to-month they are therefore stable. Both components have a positive effect as a result on the individual's security. This is, of course, the positive effect that has motivated most efforts to eliminate piecework.

Equally, since earnings only change at relatively long intervals, between rounds of bargaining at least fixed pay is unlikely to lead to disputes or stoppages. The effect on stability is therefore positive: which has been the prime motive for attempts to eliminate piecework in the UK. It has also played its part elsewhere.

On the other hand, neither of the fixed components gives the individual any direct stimulation to improve his performance or be concerned with the control of the system. Both components, therefore, get no score with regard to performance and self-correction.

The similarity breaks down when we come to the other objectives. For this reason we need to take the two components separately.

Pay for the job

Given the wage level, our concern is with how it should be made up: we do not want to get involved with the niceties of job evaluation. Nevertheless, in order to make plain the distinction between pay for the job and other components, it may be helpful to list some of the criteria by which a job may be valued. They include:

- its physical difficulty
- the nature of the physical surroundings
- its intellectual difficulty, and thus the qualifications it demands (this is not the same as the qualifications the man doing the job happens to possess. Paying more for a job because it *needs* a graduate is pay-for-the-job. Paying a man more because he has a degree is pay-for-the-man.)
- the value of the job to the firm

and several others which we might call *proprietary*, since they have been developed by the designers of specific job evaluation schemes.

One particular problem about discussing pay for the job is that it is easy to confuse it with something we want to come to later: payment for time spent working. The confusion arises because what we are calling 'pay for the job' is frequently expressed as a rate per hour, rather than a fixed sum per pay period. When there is a rate per hour, *and hours are fixed*, then, in our terms, we are talking about fixed pay for the job. When hours vary from week to week and earnings vary with them, then we are talking about the variable component, pay for time spent working.

In the most common situation there is both a fixed and a variable component. Payment is on a basis of a fixed sum for the basic number of hours, plus overtime. Fully fixed pay, where the salary is paid irrespective of hours worked, is rare in blue-collar jobs though it is common in white—though even there, of course, there is usually an understanding of what the minimum hours worked will be, and sometimes a maximum too.

The essential distinction we are drawing should be fairly clear, however. Where pay is stable from week to week, irrespective of hours actually worked (and therefore where, usually, no check is made of hours worked), we are talking about fixed pay for the job. Where it fluctuates with hours worked, we are talking about payment for time spent working.

To people from some countries, notably Scandinavia, this distinction may hardly seem worth drawing. In those countries, by and large, overtime payments and short time deductions are a relatively insignificant factor. But in other countries, notably the UK, such payments can often be a very significant factor in the wage packet, and are commonly considerable. In such countries, the drive to eliminate incentives has been accompanied by a drive to stabilize wages by reductions in overtime and the establishment of fixed 'lay-off' pay.

Apart from anything else, such a change in the pattern of payment can help eliminate class barriers. Making pay independent of hours worked places the blue-collar worker in a position more comparable to that of the white-collar man. It therefore tends to lessen the growth of resentment.

With that distinction drawn, we can look at some of the effects of pay for the job on those objectives where it differs from pay for the man.

On the economic side, the only one we did not mention in discussing fixed components generally is flexibility—the ease with which people can be shifted from job to job. It is probably obvious that, if pay is determined by the job, then it becomes more difficult to shift people around. At least, it becomes difficult to shift them from a high-paying job to a low-paying one, unless you depart from the pay-for-the-job principle, as often happens, by paying him temporarily more than the rate. Pay for the job therefore has an important negative effect on flexibility.

On the human side, there are three more objectives to look at. Group membership is barely affected. Pay for the job will make the worker identify more strongly with people doing the same job than with people, say, making the same product. But it affects only the nature of the group he identifies with, not the strength of the identification.

Paying strictly for the job gives the individual no chance to affect his short-term earnings: it is therefore scored negatively on individual influence. On the other hand, in the long-term it offers some possibility of career advancement, so it is scored positively. The effect is only weak however since that advancement will be dependent to a considerable degree on other people.

Pay for the man

Again, to help define what we mean by pay for the man as opposed to pay for the job, it may help to list some of the criteria which might be used in determining an individual's pay on this basis. They include:

- his age
- his experience
- his needs (number and kind of dependants, for instance)
- his qualifications (not those necessarily demanded by the job, but any others he may have)
- his rank (this is rare in Western industry, but is central to military payment systems, and is a fairly important factor in Japan, though rank in Japan is fairly closely correlated with age and experience).

In modern industry, at least on the blue-collar side with which we are mostly concerned, payment for the man is mostly restricted to small enterprises. There what an individual is paid may still be a result of direct agreement with his boss. It may, therefore, take into account such things as his getting married, having children, and so on.

In large companies, it has not been adopted to any significant degree. The only common exceptions are where women are paid less than men, or juveniles paid less than adults. The paper-making company we quote in the example at the end of the chapter, however, has begun paying extra increments to people with extra skills, even if they are not relevant to the job in hand.

It is much easier to think of examples from organizations like the army, or teaching and the civil service. In the army, colonels get more than majors, even where they are doing the same job. And there are usually graded allowances for special qualifications and special personal circumstances. In teaching, in the UK anyway, graduates are automatically paid more than non-graduates. And in the civil service, rank and length of service are usually of major importance.

Nevertheless, there is no theoretical reason why the potential effect of pay for the man should not be taken into account in planning industrial payments systems.

In contrast to pay for the job, that effect includes a strong positive contribution to flexibility. If a worker's pay is unaffected by the job he is doing, it is administratively much easier to change his job. His own objections will also be much lessened (though not completely eliminated—he may have objections to change unrelated to pay, connected with friendships, habits, need for job security and so on).

On the human side, the strongest contrast is on group membership. Pay for the job has a neutral effect, but pay for the man has a distinctly negative one. If a person's earnings depend solely or mostly on his own individual circumstances, it can only lessen his sense of community with his colleagues.

For the self-same reason, however, it gives him more scope for affecting his short- and long-term earnings. Both these are scored positively, though the long-term effect is much stronger, since in a pay-for-the-man system, changes can only be brought about slowly. In the short term, however, the individual retains the satisfaction of knowing his earnings are the result of his past efforts.)

Variable components

The variable components have less in common than the fixed ones did. All three have a negative effect on security, and two have the same sign on four objectives, but even here the degree of the impact varies. We need, therefore, to deal with them all separately.

Pay for time spent

We have already remarked on the danger of confusing this with pay for the job. It may, however, be worth repeating that we are talking about those elements in the income package that are proportional to time spent working.

Perhaps we should add that that means time spent 'officially' working, irrespective of output. There may be a relationship to output, in that people in a pure system of this type will not be paid if there is no work to do. It would be more accurate, however, to say that pay for time spent working is input-related rather than output-related. (The difference between these is significant in one of the examples at the end of the chapter.)

This fact accounts for the difference between this component and the next, payment for output, as far as self-correction is concerned. If the production rate were constant, and payment were only for hours worked, then unit labour costs would automatically remain constant. That would be a valuable aid to control.

But to assume that production rate remains constant is a little idealistic. The natural tendency when work is short is to slow up the work rate. And, where the situation has degenerated badly, there is a further temptation to slow up when work is plentiful, in order to generate overtime.

Such tendencies can be kept in check, but only at the expense of increasing supervision and the expense of control.

We recognize, therefore, that there is potentially a positive effect on self-correction (via unit labour costs), but we have to mark it as slight.

The total effect on performance of payment for time spent working must be taken as negative. It provides no stimulus to increase output, and in situations like that above there is a direct stimulus to decrease productivity.

It may, of course, make people willing to work longer hours, thereby increasing output. Where that enables capital equipment to be used more fully it will lead to an increase in capital productivity. It has no effect on productivity in terms of output per man-hour, however. Where the overtime rate is higher than the normal rate, as it usually is, there will, in fact, be an adverse effect on unit labour costs.

Whether that is important or not will depend on the relative local importance of labour costs and capital costs. The argument does not alter the negative score we give on performance: it simply reflects the fact that in a given situation, performance may not be the thing that needs strengthening.

On flexibility in terms of the ability to move men from one job to another, it is difficult to see that pay for time spent has any intrinsic effect. If the rate of pay is dependent on the type of work being done, then flexibility will be decreased: if it is not so dependent, flexibility will be improved. But that is only to repeat the distinction we already made between pay for the job and pay for the man: it is not affected by paying for the number of hours spent working instead of a fixed amount per week.

(It may make the system more 'flexible' in the sense that it may be easier to have people working—or at least being paid for—more or fewer hours. But here that is more what we mean by controllability than flexibility.)

It has a slight negative effect on stability, however. The need to keep accurate track of time can generate arguments and consequently disruption. But it is easier to avoid such troubles when time alone is being measured than when some function of output has to be calculated.

In most communities, too, the idea that people should only be paid for the time they work (plus holidays) is pretty deeply ingrained. In only a few places is it likely to arouse hostility.

On the human side, this component helps the individual to some extent to influence his own income. It does so, for instance, where he can choose to work overtime or not. Much obviously depends on local circumstances, and in particular the availability of work. Taken generally the effect has been marked as slight.

The effect on security is negative, as with the other variable components. But, on the whole, hours are likely to vary less than output. We have therefore given it a comparatively slighter impact.

Elsewhere on the human side, pay for time spent working has no effect worth remarking on.

Pay related to output contribution

For want of a more generally applicable phrase, we talk here of contribution to *output*. But, strictly speaking, there is no reason why it should be taken to mean volume of output only. We are also including here incentives paid for finishing to a deadline, for quality, for minimizing down time, or for keeping costs down in various ways. There is no reason indeed why they should not be linked to any economic parameter at all—or any combination of such parameters.

They should, of course, represent factors, or combinations of factors, that are critical to the success of the enterprise concerned. (This is well illustrated in one of our later examples.) An easy mistake to make is to link incentives to things that are easy to measure, rather than things that are important. To take a classical instance, salesmen's commissions may be linked to the volume of sales irrespective of their profitability. Or payments are related only to volume output irrespective of quality or cost.

But the question of determining the best measure for a particular situation is a complex one, and Maynards are not alone in devoting much development work to it. For now, all we wish to establish is that we are talking about payment for the contribution made to the firm's results—to its attainment of its key economic objectives.

The word *contribution* is particularly important. For we are concerned specifically with payments made for the part that the recipients have themselves played in attaining those results. General payments made to everyone in the firm irrespective of the parts they have played in its success are something we will be coming to later.

Nevertheless, contribution does not only mean individual contribution. This component also covers group incentive schemes, where a joint payment is made to the members of a particular work group in recognition of their joint contribution.

Such group schemes have a different effect on some objectives than indi-

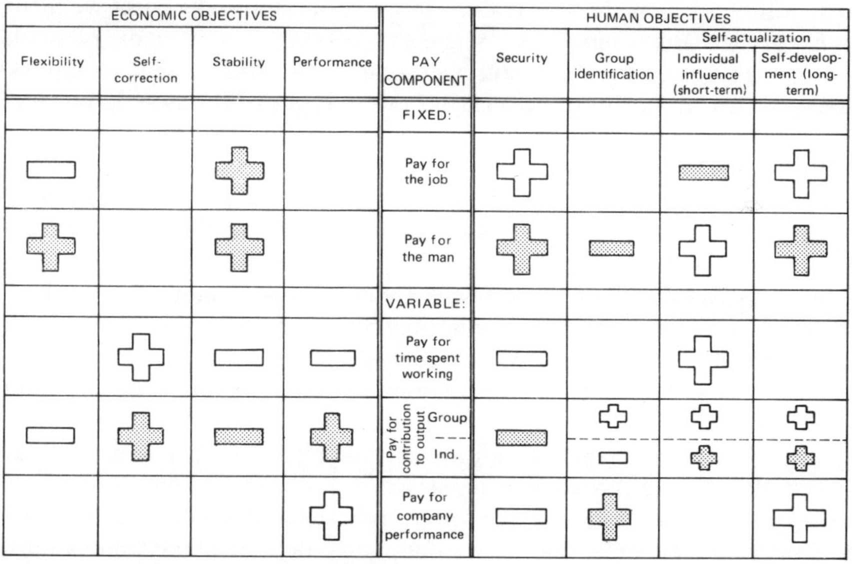

Fig. 6.2. Potential impact of pay components.

vidual schemes do. In one or two places, therefore, we have had to complicate the diagram of Fig. 6.2 to allow for the differences.

Before we look more closely at that diagram, however, it is worth pointing out again that incentive schemes have a side benefit. Installing one demands close analysis of what the employee's role really is, and what part he plays in contributing to the firm's results. Even where the scheme is a straightforward piecework one, the study required for that analysis, to ensure that standards set are accurate and fair, may be highly productive on its own account.

Unfortunately, this consideration may bias the results of studies like those we quoted from Sweden. Improvements in productivity following a transition from fixed wages to incentives may depend on the analysis involved as well as on the motivational impact of the new system.

As Fig. 6.2 indicates, that impact varies considerably from objective to objective. Economically, the strong positive impact on performance is traditional and conventional, but nonetheless real. Traditional protagonists of incentive schemes may have erred in overlooking some of the side effects. But they were not mistaken in recognizing that extra financial reward is a powerful stimulus to extra output. That stimulus may not be entirely a materialistic one. Independently of the actual purchasing power of the increase, it serves an important function as a mark of achievement and status.

Just as powerful and positive is the impact on the self-corrective value of the system. If wage costs are strictly proportionate to output, then unit labour costs will be that much stabler. Even more importantly, the effect we mentioned earlier comes into play. Where the income of individuals, or work groups, depends on the successful operation of the production system, they have a powerful motive to help ensure that the rest of the system, not just their own part, works efficiently.

Workers on incentive schemes are more likely to keep the service functions on their toes, to ensure that parts and equipment are being supplied efficiently, and to act quickly (even if it is only to notify management) when something looks like going wrong.

Indirectly, foremen and supervisors are also strongly motivated to show foresight in maintaining proper working conditions. Failure to do so will cause immediate criticism from their subordinates.

Elsewhere, unfortunately, the economic effects are negative. Once a man has developed the skills that ensure a satisfactory level of earnings in one job, he is not going to take too kindly to changing it. Without some kind of financial cushion, a cab driver who has learned to make the most out of one district by being in the right places at the right times, is not going to be eager to move to another area. The more a salesman is dependent on commissions, the less ready he will be to move to a new area when he has to build up fresh contacts. And someone who has developed high performance on one kind of machine is not going to relish having to develop a new set of skills on another.

Flexibility is therefore normally weakened by incentive schemes. Group schemes, however, do permit flexibility within the group. Stability is also weakened. Renegotiation of rates and disputes over the rates themselves and the administration of the scheme are a major cause of disruption of production.

On the human side, the picture is more favourable than it is usually presented as being. At least, it is when the whole range of objectives is taken into account.

Incentive schemes are often claimed to have a negative effect on security: the worker can never be sure how much his future earnings are going to be. Against that, in the great majority of incentive schemes, earnings do not vary more than a few per cent a month. Still, the security of incentive schemes is not equal to that of fixed wages. Incentive schemes can score positively on all the other human objectives, however.

We say 'can score' advisedly, because, in particular on the group membership objective, much depends on whether the scheme is a group one or not. Group schemes, properly introduced, foster group identification. Individual ones do the opposite, as do those misconceived group schemes that lead to intra-group discussion and frustration. One reason for union opposition to piecework, especially in the earlier days, was fear of the lessening of worker solidarity that it could involve.

Both group and individual schemes score positively for the self-actualization objectives, since they help the individual to influence both his short- and long-term earnings. Group schemes score more weakly: the influence any individual can exert is stronger if it is direct than if it has to be exerted through a group.

Payment for company performance

The simplest term for what we include under this heading would be 'bonuses', the term we used earlier. Unfortunately, though, 'bonuses' is frequently used to refer to incentive payments like those we have just been describing. In particular, it is frequently used for productivity-related payments made on top of a fixed wage. That is what we would prefer to call a 'premium', since it is a kind of incentive scheme. The payments we want to talk about here are those made annually, or perhaps twice or four times a year, and related to the overall results of the company, but not to the recipient's own contribution to those results.

In this sense, payments made under some fixed scheme for participation in profits are a 'bonus'. So too are *ex gratia* payments made irregularly and unsystematically. Such payments are becoming less common in the West than they once were, but they remain a fundamental part of the Japanese system. There the annual or semi-annual bonus may well represent a quarter or a third of the employee's total income, making his economic well-being extremely dependent on that of the company.

Schemes of co-ownership and employee shareholdings can also be included in this category. Dividends, and changes in stock market valuation, represent a kind of bonus. Apart from stock-option plans, however, such schemes are more generally intended to give the employee some share in the decision-making of the firm than to have a purely financial effect. They relate rather more closely to the subject of chapter 8.

The phrase 'participation in profits' cannot, of course, be any more precise than the word 'profits' itself. It may mean *taxable* profits. It may mean *trading* profits (thus removing the effect of the company's financial and property management). In the case of a multi-divisional company it may mean *divisional* profits, rather than corporate ones.

And there are more complex possibilities. For a while after the Second World War, the Schuller Plan, making total monthly salaries a fixed percentage of monthly turnover, was popular in France. The not dissimilar Scanlon Plan spread at about the same time from the US to Europe. Attempts have been made to make the wage bill a constant proportion of value-added.

To qualify under our heading, all such schemes have to have in common is that they are not directly linked to the individual's (or the work group's) own efforts.

Such payments are unlikely to have any direct effect on any of the economic objectives. The only one that may be influenced is performance, and then positively, but only if the individual has reason to feel that his own contribution to general profitability is significant. Profit-linked bonuses are more effective motivators for managing directors than for shopfloor workers. Any effect on performance is more likely to be the indirect result of the effect on the human objectives.

On the human side, the important fact is that such bonuses are the only component of the wage packet that helps at all with the identification of the worker with the company (rather than the work group). This is perhaps the most significant benefit the Japanese have gained. It is true, however, that, if such bonuses appear niggardly, this effect may be reversed. The individual may then be alienated from the company rather than encouraged to identify himself with it. The only safe bonuses may be substantial ones, as in Japan.

Japanese experience also indicates that the bonus system can have a beneficial effect on the self-development objective. The Western worker relies extensively nowadays on hire-purchase for acquiring the consumer durables that mark the progress of his standard of living. The Japanese, however, relies on the periodic receipt of fairly substantial capital sums. He can, therefore, avoid the continuing requirement to meet hire-purchase or other repayments which is a standard source of insecurity among Western workers. This effect, however, is not enough to overcome the general negative influence on security which is shared by all the variable components. Bonuses may not be as strongly negative as payment for time spent working, or for output. Restricting ourselves to two degrees of impact on the chart prevents us from discriminating too closely, however.

After reviewing the considerable range of effects that the different components can have, the Swedish results we quoted perhaps become easier to understand. It becomes obvious why the companies that switched to premium systems did better than those that chose fixed wages, and also better than those that stayed with piecework. Premium systems, combining payment for the job with a reasonably strong element of output incentive, take more of the objectives into account.

For, luckily, while they have different effects, none of the five components is totally incompatible with the others. Any one individual's pay packet may be made up of two, three, or even all five types of payment. With care, some combination of the five can be blended into a mix that will suit any particular circumstances. It has to be borne in mind, however, that too much subtlety may stop those involved from understanding the system. Then it has no effect. As with planning systems, over-complexity is self-defeating.

One such approach to the blending of several different components into one system is provided by our first example.

Example 1. Paper-making

This example demonstrates neatly both the complexity of the problem and the way in which different components can be blended into one package. But it needs to be pointed out that the solution is a new one, and therefore relatively untested. It is also worth noting that it was not an imposed scheme, but one worked out in collaboration with the work groups.

Like most payment schemes, the one to be replaced was itself a mixture of several components. At the heart of it was a job classification scheme by which the various crafts and skills represented in the workforce were grouped into seven grades. Each grade was paid a different rate. The major determinant of pay was therefore the job.

About 75% of the wage packet, on average, was made up of fixed earnings. Forty per cent was exactly what we have described as pay for the job. The other 35% was termed by the company a 'basic premium'. Since it was constant, however, and varied between job grades, from our point of view it could also be called pay for the job.

The variable part of the package was made up of two major and one minor component. Of the major ones, the first was a premium related directly to volume output (in tons per day). The other was paid for shift working, and also represents an output-linked component, though related to continuity of output rather than volume.

The minor variable component was represented by extra pay for Sunday and holiday working—an example of pay for time spent working.

As a premium system (in the language of the Swedish study we reported at the beginning of the chapter) it might be expected to work reasonably well. It would have neither the disadvantages of fixed-wage systems nor those of pure piecework.

Indeed, from the point of view of the objectives of Fig. 6.1, it did work reasonably well. But comparing the system with that chart indicates two areas of potential weakness. It does nothing for flexibility, and little for self-actualization.

From the chart, both of these could be corrected by building in an element of pay for the man, and this was at the heart of what was in fact done.

The previous seven grades, each containing its specific skills and crafts, were replaced by four. But instead of simply being numbered, like the previous grades, the new ones were given titles. In reverse order of seniority, they were: paper maker; senior paper maker; paper technician; senior paper technician.

Naming the grades may seem a minor change. But it succeeded in focusing attention on membership of the group (of paper makers, etc.) rather than the craft, as well as on status and rank inside the company. (New recruits spend a month in training before becoming 'paper makers'.) Group identification was therefore helped, countering the tendency to weaken it which is inherent in the 'pay-for-the-man' concept which was also to be introduced.

Within each grade, basic pay is fixed and common. However, there is also a moderately substantial experience increment, paid as follows:

Group	*Increments paid after*		
Paper makers	6 months	1 year	2 years
Senior paper makers	6 months	18 months	$2\frac{1}{2}$ years
Paper technicians	6 months	18 months	$2\frac{1}{2}$ years
Senior paper technicians	6 months	18 months	$3\frac{1}{2}$ years

No more than three service increments are paid in any grade.

However, it was also made possible for anyone to improve his fixed pay without moving up a grade. This is through the payment of a series of fixed allowances for extra skills. The details of the system are too closely tied to paper-making technology to be worth giving. But its essence is that a machine minder, say, who is also qualified as a dryer, receives a permanent fixed increment for the qualification. He receives that payment, not for working as a dryer, nor for *doing* both jobs, but simply for being *able* to do both.

In the two lower grades, four such increments are possible. Anything more than that would be likely to lead to promotion anyway. In the technician grade there are three. In the senior technician grade there is only one. By the time anyone has worked his way up through the grades there are not many new skills to acquire anyway.

The system is an obvious direct inducement to flexibility of manning. It strengthens still further the natural advantage of pay for the man in this respect. However, it also has a powerful effect on both the self-actualization objectives. It provides an extra way for the individual to influence his earnings in the short term, and it holds out the prospect of long-term career development.

Finally, on top of all this, there is a variable, and strictly output-dependent premium, group-based, and linked much as the previous one to tons processed per day. However, instead of being strictly dependent on the output in the previous pay-period, it is tied to a three-month moving average. A positive performance incentive is thus maintained while insecurity is lessened.

As we said, the system still has to prove itself. Since so many of the advantages it offers are long-term, it may take some time to do so. But it still serves as an example of how one system can take into account a multiplicity of objectives.

Example 2. Engineering industry

Simplicity too can have its merits, as is shown by this example from a Danish metal-working factory with about 800 employees. Previously, it had had a fairly traditional piecework scheme. But, with something like 80 000 different operations to be taken into account, incentives were loose and unreliable, being based on inadequate and poorly organized work standards.

Moreover, the wage level in the factory was lower than the statistical average for the industry. That resulted in a union demand for a 10% wage increase, and the renegotiation of new work rates.

Simply to do that, without a comprehensive programme for replacing the existing work standards, would have brought no lasting benefit. On the other hand, the large number of different operations involved would have made such a replacement a lengthy project, and taken comparatively large resources. Increased productivity might have justified the cost, but management decided there was not enough time available. Instead, they accepted a consultant's proposal for a simplified scheme—'multi-level wages'.

A quick (though admittedly not thorough) analysis of the workshops revealed ways in which productivity might be improved. Mainly achievable through reducing the proportion of indirect to direct labour, and making the service functions more efficient, these improvements were utilized to compensate for the wage increase asked for. The old standards were adjusted to allow for the change.

Piecework incentives were then dropped. Instead, pay was linked directly to the attainment of the potential productivity increase. 70 to 80% of total wages was paid as a fixed sum (based on the job). Three productivity levels were then set, on a group basis. Group productivity was computed as a moving average over three pay periods. As each level was attained, a specific increment was added to pay, remaining fixed until either a new threshold was reached, or productivity fell back again below the previous one.

In fact, the second of these two alternatives never happened. Over 18 months, productivity steadily improved to a level equivalent to an increase in capacity of 24%. Wages for skilled workers increased from 14·80 Danish crowns per hour to 17·22, and for unskilled workers from 13·6 to 16·19. Thus the workers got more than the 10% they had sought, but improved productivity more than compensated.

It is true that the productivity increase was partly due to the improvement of working conditions. It was not a result of improved performance alone. But, equally, without the direct linkage of rewards to objective achievement brought about by the new wage system it is unlikely that the potential revealed by the study would have been gained in practice. Or, at least, that it would have been realized so easily and so swiftly.

Example 3. Timber

Our final example, from a sawmill with 50 employees, deals only with the replacement of one kind of output-related premium by another. But it is a neat example of the need to tie such premiums to meaningful economic objectives, if they are to function properly. Moreover, both the old and new systems were tied to the work of the group rather than the individual, but the ways in which they were linked differed significantly.

Originally, the premium depended basically on the number of logs sawn, and their diameter. Different rates were paid for logs up to 5¾ in., 6¾ in., 7¾ in., and so on. Other increments were also paid. They could be received for different kinds of supplementary work, for working in winter conditions, for changing from one log consignment to another, for changing saw blades, and so on. Down times of up to 30 minutes were ignored, but above that a time rate was paid.

All of these rates were set in 'per man' terms. Thus, the average employee's piecework rate would be set at *x* öre per 100 logs sawn by the group, irrespective of the number of men in the group. (Differently skilled workers were paid different rates, but, from the point of view of this example, that is irrelevant.)

On its own, each of these components had its rationale. In combination however, they led in the wrong direction on several counts.

In the first place, it was not output that was being rewarded, but input. The workers were being paid, not for the amount of usable timber produced, but for the amount of work they did and the amount of raw material they used to produce it. Additionally, they were paid extra for down time that lasted longer than about 30 minutes. Effectively, therefore, they were rewarded for making down time last that long. There certainly could be no question of their being eager to get back to work after 25 minutes.

In the same way, there was no incentive to save time in replacing saw blades or in changing from one consignment of logs to another. And, finally, since the amount paid to each man was directly tied to the output of the work group, there was every incentive to try and increase the group's size. (It may not be obvious at first sight, but the effect of the arrangement outlined above was that adding one man of average competence to a group of ten would increase everyone's income by 10%.)

The objectives that the system was really tied to could therefore be expressed as follows: 'Use as much raw material as possible. Ensure that down time is never less than 30 minutes. Don't worry about wasting time over changing consignments or replacing saw blades. Make sure as many people are employed as possible.'

That probably was not what the original designers of the system intended. But it was what they produced. And it needed changing.

From the management point of view, it needed changing to something like: 'Produce the largest possible quantity of approved sawn timber per man-hour, including time for consignment and saw blade changes, and other down time.' However, the amount of effort the worker has to exert to produce a given amount of timber depends on the kind and quality of logs he is presented with. As bald a statement of the objective as that would therefore have been unacceptable. It was necessary to take into account in determining pay the amount of work done, i.e., the amount of timber sawn.

Once the objective had been clearly defined, it was relatively easy to see how the premium should be designed. It has to have two components: so much for each log sawn, and so much for each cubic metre of finished and approved timber. The payments for changes and for other down time had to be eliminated. And though the premium was still required to be group-linked, it needed to be related in a way that took account of the size of the group.

A group rate of x öre per log (irrespective of diameter) was therefore set. So too was another, of y öre per cubic metre of acceptable output. Once the group premium had been calculated, it was then divided between the members of the group. Increasing the size of the group, while it might increase the total premium, therefore made no difference to any individual's share.

From a technical point of view, it is worth noting that introducing a true output component made it unnecessary to take into account variations in log diameter. The original variations were introduced to take into account the fact that you get more timber more easily from a large log than from small ones. Adding in the direct measurement of output made this unnecessary.

It no longer made much difference from the worker's point of view either. With small logs going through the plant, the output component of his premium might be lower, but the input component would be higher. With large logs, the situation reversed itself.

This last point is of more general application than it may seem. In many situations complex indirect ways of measuring output have evolved historically that could well be replaced by much simpler direct ones.

7. Housing the system

The architect and the production designer. Internal environment: sound, light, air, heat and layout. The factory and the community. Dust in a foundry—amenities in a brewery.

Whose design?

If, in opening the last chapter, we felt a need to justify the production system designer concerning himself with pay schemes, the feeling is even stronger here. Traditionally, designing the building to house the production system (where it did not have to be fitted into an existing one) was a matter for the specialist, the architect. In particular cases, in highly automated process production, or where products, machinery or other equipment were abnormally massive or powerful, he might have to meet particular specifications. He would undoubtedly also be subject to a cost constraint, but, over the run of industry, designing the shell to put around the machinery was the architect's business.

As it still is, by and large. Since the need for special knowledge and qualifications is indisputable, the architect is indispensable. But his function cannot be regarded as independent of other aspects. For, just as the pay subsystem interacts with other sub-systems in a way that may destroy their potential, so, too, does the physical environment in which the work takes place.

To some extent, it does so directly and visibly. At the most elementary level, a wrongly designed building may prevent the installation of appropriate production facilities. Or it may provide inadequate space for holding stocks of materials or components, thus creating problems of utilization, and hindering performance. Thus it can weaken two of the three components of productivity.

It would be fair to say, however, that this problem, though it exists in theory, is not a serious one in practice (except occasionally in the abnormal circumstances we mentioned earlier). Generations of experts have applied their talents to the solution of what is a continually recurring problem, and, not surprisingly, they have approximated fairly closely to an ideal solution. Hence the evolution, over the last few decades, of what has been called the 'multi-purpose box'. A huge industrial hall on one level, without partitions, with a few pillars and preferably no windows has become the archetype of industrial building in modern times, wherever it can be fitted into site constraints.

Whatever its aesthetic merits, such a solution would not have become standardized if it had not helped to achieve each of the three objectives virtually every designer was attempting to attain: high productive capacity, flexibility, and economy. If those were the only objectives that mattered, the production system designer could well leave the whole business to his professional colleague.

But they are not. And once one begins to take human objectives into account, the picture changes.

To some extent, the effects on the human objectives are tangible and measurable. In particular, this is true of the effect on security in that aspect which is to do with physical safety and health. The effects of noise on hearing, of heat (and cold) and atmospheric pollution on health, either directly or through long exposure, are all relatively easy to establish. There may be technical problems involved, but nonetheless they are limited and well-defined problems that can be left to the specialist.

Other effects are much less tangible. In the extreme, people's perceptions of even the physical environment can differ substantially from reality. In one fairly recent sound simulation experiment carried out on industrial premises, a tape recorder was used to play back the sound of ventilation fans. The volume was gradually increased in order to study the reaction of the people present to the noise level. Those conducting the experiment registered some slight surprise when, towards the end, some of the subjects began to complain not of the noise, but of the cold draught.

But, even leaving such distortions out of account, the effect, say, of building layout on workers' relationship to their work group, is highly subjective. In any case, the importance or otherwise of such effects on the whole system is closely related to the importance that the production system designer wishes to give to those particular objectives. If group membership has been discounted as an objective, it may not matter very much if people can talk to one another, or not. On the other hand, if it is considered important, the ability to communicate easily may be vital.

On the human objectives, therefore, cooperation between the production system designer and the architect is not only useful, but essential. (Indeed, in view of the subjectivity of people's responses to environmental conditions, there is a strong case for involving them in anything that changes those conditions. But the subject of participation in change is one that we are reserving for chapter 10.)

Other specialists, too—physiologists, psychologists, various medical specialists—are also likely to be involved from time to time. But knowing when and how their specialities are likely to be useful must become a part of the system designer's job. In the context of a book like this, therefore, at least a brief consideration of the various elements that go to make up the physical environment is essential.

Internal environment

That environment can be described in terms of five basic components, sound, light, air, temperature and layout. All of them affect to some degree the human objectives of the people working in the environment. And all of them are subject to technical modification. Sound can be muffled and absorbed. Light intensity, contrast and colour temperature can all be varied. Atmospheric pollution can be controlled. The temperature of the workplace can be regulated. Whether they should be, and how far they should be (and at what cost), are not such easy questions to answer. To help in understanding the difficulties involved, we will take a brief glance at each of the components in turn.

Sound

For the most part, the problem of noise in industry originates in the production process. However the technical support system of the building itself—compressors, generators, ventilation systems, and so forth—often contributes its own share of noise. Inadequate screening and insulation plus the use of hard materials in walls and ceilings distributes noise throughout the building so that even quiet places of work suffer from disturbance. But the problem is obviously most serious in close proximity to machinery, where the high sound level can be physically damaging as well as leading to social isolation.

Here one can with some objectivity assert an absolute minimum objective: that workers should on no account be exposed to sound levels that can result in defective hearing. At our present level of understanding of the processes involved, that objective can be quantified as a maximum noise level of 85 dB for a maximum of five hours a day.

Beyond that point, however, more subjective considerations come into play. From the point of view of maintaining performance, let alone considering the effect of the worker's comfort, the aim should be that the noise level should not be psychologically disturbing, or make it difficult to concentrate, or lead to irritation or exhaustion. If that is to be achieved, a sound level of much less than 85 dB, or a system for rotating workers so that they do not spend too long at noisy workplaces, is going to be needed. If the latter, then the work planning system, and possibly the production mode, will be affected.

Moreover, as we have already pointed out, if one of the desired objectives is the establishment of a major degree of group identification, then the noise level will have to be low enough so that people can talk to one another easily. Or, again, the production mode will have to be chosen so that jobs can be rotated. Or there will have to be provision made in the layout of the building for areas where people can confer with one another in reasonable quiet, and provision made in work planning for them to use them.

Noise *level* is not the end of the story. Noise at particular frequencies can have disturbing effects greater than would be accounted for by their decibel level alone. The reverberation time (the 'echo effect') of the premises can also produce deleterious effects.

Finally, there is plenty of ground for believing that, for some people at least, the occurrence of sudden, unexpected noises can produce stress. People's susceptibility on this count varies widely, however, and it remains an area where we need to know more about the physiological and psychological mechanisms involved before we can do more than treat the matter with caution.

Light

Bad lighting conditions are usually the result of incorrect planning of artificial lighting in conjunction with daylight conditions. Sunlight itself is inherently contrasty, producing bright highlights and dark shadows. On the other hand, artificial lighting is normally installed in a stereotyped way, dependent on suspended fluorescent tubes and yielding an even, undifferentiated light.

Simply increasing the strength of such a system, especially when it is coupled with the elimination of windows for the sake of building economy, achieves only a brighter, still even, light. What is required for maximum performance, as well as worker comfort, is a fairly high degree of variation in lighting, matched to the requirements of the production layout.

Adequate variation in light and shadow is necessary to ease the strain of estimating distance and discriminating between shapes. And, in general, for efficiency and comfort, the worker needs the light concentrated on the place where he is working, with peripheral areas given less illumination. To achieve this, collaboration between the architect and the production system designer is obviously necessary.

But, for ultimate flexibility, the simplest solution is probably to provide the worker with adjustable light fittings. This not only helps to ensure that his light conditions are correct for any particular task. It also must have at least a minor effect on his ability to influence his own circumstances, and hence on his self-actualization needs.

Not only the intensity of light, but also its colour is of some importance. So the colour scheme of the surroundings has to be taken into account.

Partly, this is a physiological phenomenon. The eye is most sensitive to light in the yellow-green part of the spectrum. Sometimes it is possible to put this fact to good use, as Kodak originally did in their laboratories at Rochester, where the introduction of very narrow-band yellow-green lighting (to which even colour films and paper are relatively insensitive) enabled darkroom lighting conditions to be considerably improved.

More usually, however, the effects are psychological and intangible. Since colour schemes tend to be of little economic importance usually, they provide a rather obvious way of giving the worker some freedom of choice, with presumably helpful effects. (However, this may have been rather overdone in the past. It is of such obvious economic unimportance that workers may well feel it is being used as a sop to human relations.)

Air

Air pollution in manufacturing industry (with some exotic exceptions) is caused mainly by welding fumes, grinding dust, oil mist and gases. Where the source is static, it presents little problem from a technical point of view. As long as the need for it is foreseen, it is usually possible to provide efficient extraction and ventilation.

There are more serious technical problems in the case of mobile equipment for, for instance, welding and grinding, when it can sometimes be very difficult to arrange for effective mobile devices for extraction. This may be avoided by proper attention to the problem in designing the production layout—by taking the material to the grinder, for instance, instead of the grinder to the material. But that is not always possible. There will always be a trade-off here between the technical difficulty (and cost) on the production side, and the technical difficulty (and cost) on the building side that will certainly not be effectively solved if the architect and the production system designer work independently of each other.

Temperature

Temperature control and air control have, of course, much in common, since proper design of ventilation and air conditioning systems do much to solve both problems. As an objective, however, controlling temperature (especially through heating, but also through cooling) has been part of the architect's task longer than any other of the components we have so far dealt with (it is 50% of the reason for having a building in the first place, the other 50% being to avoid the weather).

As a result, only carelessness or poor design work can, over the general run of industry, account for failure to provide reasonable temperatures for working. Even in those industrial processes where high or low temperatures are an essential factor, counter-measures to protect workers are well developed, though problems remain in, for example, welding in shipyards during the winter. Which is not to say that cheaper or more economical measures for doing the same job cannot or will not be found. But that *is* a job for the technical specialist, and from our point of view barely worth commenting on, since achieving the same results more economically is not going to clash with either the economic or human objectives of the production system.

Layout

Admittedly, as we have said, from the point of view of the economic objectives, the now-classic 'multi-purpose box' in which anything can be moved anywhere at any time is probably as effective a solution as can be achieved to the layout problem. But, as we have already hinted in talking about noise, taking human objectives into account considerably weakens its validity.

In particular, of course, the degree and kind of social contact among the workers, and the structure of the groups that are formed, will very largely be determined by the physical restrictions of the buildings, including such things as local noise levels as well as the positioning of screens and partitions. Where a deliberate attempt is being made to foster some particular kind of group structure, physical arrangements need to be planned with that attempt in mind.

The production system designer may try to create a work planning system that depends on collaboration and cooperation within and between groups. It is doomed to failure if the physical layout of the building prevents such groups from forming—or encourages the formation of other groups, with their own internal and external relationships, loyalties and allegiances, with no connection with the logic of the production process.

It may be worth pointing out that this requirement means neither that screening is good and open-plan systems bad, nor the reverse (though both claims are sometimes made). It simply implies that the amount of screening and partitioning needs to be deliberately planned, taking into account (as with everything else we discuss in this book) both economic and human requirements.

External environment

So far, we have been concerned exclusively with the environment inside the plant. It is worth making at least a short reference, however, to the external environment, in the sense of the location of the plant.

Traditionally, community planners have taken it as axiomatic that different areas should be reserved for different activities. Commercial, residential and industrial building are all given their own zones. As far as the segregation of industry is concerned, this partly arises from the advantages of being able to make special provision for the various community services (especially roads for heavy traffic) that industry needs. It has also been a reflection of the feeling that the presence of industry disrupts residential areas simply because factories are unpleasant places to live near. That may once have been true: with regard to some industries, it may still be true. But for most industry it is certainly exaggerated. And there is much to be said for at least a partial reversal of the trend.

From our immediate point of view, the major consideration is probably that it is much easier to generate group spirit in any workforce if the enterprise that employs it is felt to be an integral part of a rounded community. The achievement of many of the objectives we have referred to in previous chapters—group membership, product identification, self-actualization (through the possibility of encouraging community work, for instance)—is therefore made much easier.

There are other arguments from the point of view of the community at large which also counter the basic objection that this kind of development is harder to plan. For instance, it is fairly well established that children who grow up in onesided residential suburbs are deprived of many essential experiences. The categorical splitting up of a town into areas having different functions also destroys qualities concerned with well-being and 'homeliness'. It creates anonymity, and inhibits contact between people. And mixed development makes, on the whole, for a considerable shortening of the time needed to get to and from work—an element that continually takes up more and more of what should be leisure time.

To achieve it, however, means establishing that the original feeling that led to zoning is wrong. Not all factories are unpleasant places to live by, though some necessarily are, because of size or the technology involved. In present circumstances, at least, the unpleasant proportion is a shrinking one, partly due to pressure for environmental control. Industrial architecture should strive for aesthetic appeal that has been lacking in the past.

It is difficult to demonstrate that the aesthetic appeal of a building has any relationship to the efficiency of the process that goes on inside it: or even to satisfaction of the human objectives of the people who work in it. But it will certainly be an important element in any attempt to reforge the links between industry and the community it serves.

To go any further down this track would lead us too far from the subject we have set ourselves. Instead, therefore, let us look at two recent examples of attempts to find a new solution to the problems of the physical environment of the production system.

Example 1. Car industry

For the specific problem of dust-pollution, one of the most pernicious of manufacturing processes is that of fettling—flash removal—in a foundry. It tends to be a noisy job, and involves heavy lifting, but the chief difficulty is that the process of cutting or grinding off the waste material left on castings by the moulding process is inherently a prolific source of metallic dust and considerable noise.

Over the last two years, however, the Swedish factory of this example has carried out a nearly $2 million programme of re-equipment and redesign of the department's work with the specific aim of improving environmental conditions, and with a considerable degree of success.

Inherent in the programme from the beginning was a high degree of participation from employees. Consultation groups were formed of workers and supervisors together with technicians and ventilation and other specialists, to deal with everything from the installation of hoisting gear and tool racks to lighting and colour schemes and the key questions of evacuation and ventilation.

As a result of the programme, ten one-man and four two-man grinding 'berths' have been installed in the department. Each is built, apart from windows, of sound absorbent material so that outside the berths people can talk together normally. Even inside the berths the noise level has been reduced from well over 80 dB (A) to a more bearable 75.

Lighting in each berth has been increased to 1000 lux—four times stronger than before—with the installation of four fluorescent tubes. These are deliberately placed so as to eliminate shadows on the castings being worked on, though each operative is also supplied with an inspection lamp, mainly for inspection inside castings.

For grinding, each casting is placed on a worktable with a built-in evacuation and ventilation system. The operative is also supplied with a vacuum hose that he can direct as required. When working inside a casting, the hose can be placed inside, ensuring that no dust can come out of it. The ventilation system changes the air at a rate of 5000 cubic metres an hour: an increase of 67% over the previous system. In milligrams per cubic metre, the dust content of the air now measures 1·0: before it was nearer 6·0.

The problems involved in cutting were dealt with in a different way, mainly through the development of automatic machinery. The entire machine is housed in an insulated, glass-fronted box, with the controls outside. The casting is manually loaded into a holder: when the machine is started, it is automatically transported into the box for cutting. That process finished, the holder is returned to the outside for reloading, while the waste material is automatically taken away on a conveyor belt.

As a result of the innovation, the dust content of the air, previously slightly worse than that in the grinding area, has been reduced to roughly the same level of 1·0 milligrams per cubic metre. And noise level, previously a danger level of 99 dB, has been reduced to a safe, though admittedly still noisy, 80 dB.

Example 2. Brewery

Conditions inside a modern brewery are vastly different from those inside a foundry. The problems confronting the designers of a new 130 million litres a year plant in Sweden were therefore very different from those described above.

Of the two physical factors involved, noise and temperature, the first was controlled partly by ensuring that the building was soundproofed throughout, but it was also helped by making low noise level a major purchasing specification in selecting and ordering the new machinery required.

Temperature is of importance because at various stages in processing the liquids involved have to be stored at unacceptably low temperatures. Luckily, however, the process is a continuous one. The problem was therefore circumvented by locating all storage tanks in a building separated from the main one, and connected to it by pipelines. No-one is therefore required (at least as a matter of routine) to work in the low-temperature areas.

Great emphasis was also placed on the design and provision of physical facilities to foster a suitable social environment, and on obtaining the close involvement of production management and workers in the whole design process. As a result, the new plant contains several features felt to be of importance in promoting the cohesion of the workforce.

For instance, all employees enter the building through the same entrance. This leads directly into a so-called 'indoor street'. This has three levels of peripheral areas—changing rooms, lecture rooms, offices, the canteen and a club room—connected by elevators and open staircases, and equipped with newsstands and vending machines. The 'street', which runs east–west, also connects with all the production facilities, all of which are on the south side.

As well as the communal areas of the 'street', all the work areas are also equipped with restrooms, each with armchairs, telephone, music/radio installations and hot and cold drink-vending machines. It is interesting to note that the decision to have a series of small restrooms of this type, each connected to a particular working area, was made by one of the consultation groups of workers formed as part of the participation programme. They preferred it to the suggested alternative of one large, communal restroom for the whole site.

Followers of Herzberg will, of course, point out that such installations affect only 'hygiene' and not 'motivation'. And it is true that we may seem to be moving away again from the essentials of the production system. Yet this example demonstrates the foresight we need to display and the length to which we may have to go in an increasingly affluent world in order to continue to attract competent staff over the decades for which the building may stand.

8. Organizing for participation

Legislation, ethics, economics. Who wants to participate? About what, by whom, how much, and how? The parallel hierarchy. Some warnings. The need for common interests.

In chapters 4 and 5, we were concerned with the degree to which the worker can share in the responsibilities of production and work planning. We were there only interested in his involvement in decisions directly affecting his day-to-day work. We restricted ourselves to matters of local importance and short-term consequence.

In chapters 6 and 7, we broadened the picture to some extent. We introduced the question of profit sharing, for instance, and the workers' involvement in decisions affecting long-term investment in plant and buildings. But on the one hand we were specifically concerned with the worker's pay-packet, and on the other with his immediate physical environment. Self-evidently, we were concerned with things that directly affected him.

In all of these areas, the extent to which the worker should be involved in decision-making may be an open one. But even the most hardened authoritarian could hardly deny that they were matters with which the worker was concerned.

In this chapter, however, we need to go further.

Legislation, ethics and economics

Changing attitudes to the role of the worker in industry have raised the question of his involvement in matters far beyond those that immediately affect him. It has happened in different ways in different countries. But, more and more, in practice, and not only in theory, the worker is beginning to be involved in areas of planning and decision-making that once upon a time would automatically and emphatically have been considered none of his business. He is being consulted in areas where once upon a time he would not have been—and in areas where it would once have been thought hardly necessary even to inform him of the decision afterwards.

In a growing number of countries, this movement has gained legislative backing. It has not done so in the United States, nor yet in the United Kingdom—though the UK, under pressure to conform with the rest of the EEC, is teetering on the brink. In Italy and Scandinavia there is no specific legislation, but national collective agreements provide an effective substitute. Countries with no requirement for some representation of workers in the higher decision-making of the firm are probably today in a minority.

As a result, the variety of legislative provisions is much too great to tabulate in a book of this kind and length. The table at the end of the chapter shown in Fig. 8.3 (page 130), however, provides some examples from a range of Western European countries.

Quite apart from government action, the feeling that it is ethically right for the worker to have a share in corporate decision-making has also grown considerably. The belief has produced its own crop of participative organizations. Many of them have a long history. The first cooperatives, and the New Lanark cotton mills of Robert Owen, for example, go back to the pre-Marxist days of the early nineteenth century.

In such cases, the motive for introducing participation is quite simply that it is right to do so. In the extreme, it leads to vesting ownership in the workers. One of the best-known examples—and the most successful—of such extreme action is the John Lewis Partnership in the UK, a countrywide chain of department stores.

There, all workers are automatically members of the Partnership. In turn, the Partnership owns all the shares in the trading company, a device which avoids all the technical problems involved in giving shares to employees individually. (Shares normally can be traded, or kept when the employee leaves the company, in a way that membership of a partnership cannot be.)

But the setting up of the Partnership in this case depended on the fact that the man whose idea it was, John Spendan Lewis, was himself sole owner of the company. In general, if participation is to be extended simply on the grounds that it is correct to do so—morally—then, in the capitalist system, it is difficult to argue that it can be done by anyone except the owner(s).

Normally, however, management is divorced from ownership. The case that workers should participate *even if it lowers the quality of decision-making*, is therefore one that management is hardly in a position to argue.

However, the argument is often advanced that increasing participation holds considerable potential for increasing productivity. Roughly, the argument runs that increasing participation on the one hand improves decision-making, and on the other lessens resistance to change. And via both those routes, it results in gains in productivity. If the argument is valid, then management does have a case to consider and present.

Briefly, then, there are three main strands in the movement for increased participation: the legislative, the ethical, and the economic. And, matching those, there are potentially three motives for a management wishing to install a system that will provide increased participation. It may be because they want to comply with the law. It may be because they want to satisfy their own social conscience. Or it may be to improve results.

To the manager with the first of these motives, the only advice that can be offered is legalistic. One can only tell him what the legal requirements are in any country in which he wishes to operate, and perhaps estimate what the future trend of legislation is likely to be. While such advice is practical and useful, this is not the best place to offer it.

It also is not easy to offer advice to the manager with the second, purely ethical, motive. Whether increased participation is of itself right or wrong is a matter for the individual conscience. And such matters are best left to a priest, a philosopher, or perhaps a politician. It is not something we wish to comment on.

On the third, economic, motive, it *is* possible to comment objectively. It is not possible, however, simply to say without qualification that such and such a system is always good, or always useful. In general, blanket advice of that kind is out of the question; though one can say that, law and ethics apart, *no* form of participative organization is worth installing unless a clear and practical *use* for it can be foreseen.

That it makes the manager feel that he is moving in the right direction, or being contemporary and fashionable, is *not* such a use.

Who wants to participate?

A great deal depends on local circumstances, and, in particular, on the attitudes and beliefs of the employees involved, at all levels. The most naive of mistakes is to assume that everyone *wants* to participate, even in decisions that directly affect them. If that were so, it would not be necessary for countries like Australia to make voting in government elections compulsory. And polling percentages would not be so low in countries where it is not compulsory.

Generally, it is probably true that the higher ones goes in the hierarchy, the greater the proportion of people who want to participate. After all, traditionally the only outlet for a desire to participate has been through promotion. But at most levels there will still be a significant proportion of people with no interest in playing a part in the affairs of the enterprise outside their own immediate section. There will be others who doubt their own capacity to do so. And there will usually be a number who have no doubt of their own capacity, but prefer not to exercise it. For some it is pleasant to be able to say, afterwards, 'if only they'd asked me . . .', but not so pleasant actually to be asked.

A prerequisite for the development of any participative system must therefore be a reasonable understanding of the people who are going to be working in it. Given that, there are three things we can do. We can describe various ways of achieving increased participation. We can attempt to estimate the potential for good or ill of each way. And we can provide a few generally (though possibly not universally) valid rules of thumb for avoiding one dangerous trap: the substitution of expensive contemporary errors for costly oldfashioned ones.

We will do all three. But first we need to revive the distinction we made in chapter 5 on work planning. Participation and delegation are different things, and they have different effects.

Indeed, participation can quite easily work against delegation. Most government legislation to increase participation in fact does just that. In particular, the introduction of worker representatives on the boards of companies is essentially a centralizing factor. Its only result must be to increase the influence and authority of the central organization. (If it does not, it would seem rather pointless.)

The same is true of the granting of statutory authority to works councils and the like. The quicker reaction times and more accurate local decisions that one expects to accompany sizeable delegation are impaired just as much, if not more, by the need to refer to a works council as they are by the need to refer to a hierarchical superior.

Conversely, the traditional hierarchy to some extent encourages delegation. It is based on single decision-makers. And it quite quickly becomes evident that a single decision-maker does not command all the necessary skills and knowledge: he has to delegate. The same is not true of a group. The group more easily assumes that it has the necessary expertise, and is thus more easily tempted to retain authority. A group is also more tempted to rely on voting instead of reasoning.

The danger that increasing participation may lead to over-centralization and excessive reliance on vote-counting is therefore something that will always need to be guarded against.

Key factors for participative systems

With that said, in order to describe ways in which participation can be handled and to form our own policies and systems for dealing with it, we need some kind of framework. That we can provide by identifying four factors which between them can specify any particular participative system. These are its *scope*, its *selectivity*, the *degree of involvement* it envisages, and its *formality*. We can look at the effect of each of these in turn.

Scope

By *scope* we mean the kind and number of issues on which participation is practised to any degree. Even in countries with comprehensive legislation, it is rare for all issues to need decision by the participative mechanism. In full worker cooperatives, all issues may in theory be subject to participative decision; but the practical demands of the situation normally make it unlikely that the theory will be carried into practice. Similarly, the theory of shareholders' control which is central to traditional capitalism does not normally allow shareholders actually to participate in much decision-making.

In practical systems, therefore, the scope of participation is likely to be limited. The criteria for limitation are likely to be connected with the time element and the need for specialist knowledge. The longer-term the decision, the more widespread its effects, and the more general the inputs of judgement, experience, knowledge and skill need to be, the more likely participation is to increase the quality of the decision-making.

Thus, long-term decisions about investment in, say, arms manufacture, or in countries with undesirable political systems, would seem to be eminently right for participation. But there can be little call for anyone but a specialist to be involved in deciding the short-term investment of surplus cash.

Selectivity

The selectivity of the system is a matter of who participates, and why. There are several possible principles that can be invoked, but for practical purposes they boil down to three: power, involvement and qualification. Of them, the most basic is probably participation on the basis of *power*.

Under this principle, those parties are allowed to participate who have sufficient power or influence to affect the degree to which the decision can be successfully implemented. This principle underlies most traditional systems of collective bargaining, as it does big-power diplomacy. Management and unions meet face-to-face, formally or informally, to settle the issues where the dissent of either party would make any solution unworkable. In theory, of course, this could mean any issue. In practice, there are likely to be issues on which unions are neutral, or which they consider irrelevant to their interest.

Such participation, successful or unsuccessful, is in fact inescapable. If a decision depends for its success on someone's agreement, then it is pointless to attempt to bar him from participation. All one can say is that it is better to recognize the fact rather than try and hide it (to save face, for instance), and that participation should be early rather than late.

The second common principle is that of *involvement*. Under this, everyone should be allowed to participate in a decision who is affected by it. If the first principle can in some sense be termed the 'realistic' one, this is the 'idealistic' one.

Unfortunately it is not as clear-cut as it seems. 'Affected by' is too vague a term. In the extreme, it can always be interpreted to mean anyone employed by the company, in which case it degenerates into participation on the basis of membership. We use 'degenerates' advisedly, since it must be obvious that any system that calls for full participation by everybody on everything must end in futility.

On the other hand, where someone is directly and intimately affected by a decision, it is difficult to deny some degree of participation—if it is wanted.

As always when intangible things and matters of degree are involved, it is impossible to be precise. All one can do is to state that the more directly and significantly someone is involved in a decision, the more his voice should be heard, and the more weight it should carry.

It is worth noting that strict application of this principle may reduce the notional amount of participation in the system. The decision should, in this case, be restricted to those most involved, without interference from larger groups as well as from superior individuals. Thus, if a group of workers are involved with the relevant specialists and their immediate management in planning, say, the layout of a workshop, there is no more reason for a full works council to be involved than for higher management. For that matter, there is no reason, under this principle, for a trades union to be involved, unless the workers feel they would prefer representation rather than direct involvement.

Other than power and involvement, the third principle of selectivity is that participation should be based on *qualification*. That is, the right to participate should depend on the possession of the relevant skills and knowledge.

Where the aim of the first principle was to recognize the realities of the situation (the political imperative), and the second to satisfy moral requirements (the ethical imperative), the aim of this principle is simply to improve the quality of decision-making (the efficiency imperative).

As such, it is the closest of the three to our personal purposes. And it is relatively easy to state the argument simply. In the traditional hierarchical situation, decision-making power is based on rank and appointment. In the conventional 'democratic' participative systems it tends to be based on membership or election. But neither basis is in fact adequate to ensure that the right combination of skills and knowledge is being brought to bear on decisions. That can only be achieved by a system in which those participating in any particular decision are those best qualified to deal with the specific matter in hand.

That is far from easy to achieve, and perhaps any kind of formal system militates against it. But it ought to remain the major goal of anyone attempting to design such a system.

Degree of involvement

Having established the three main principles of selectivity, we can move on to discuss the varying *degrees* of participation that are possible. While we use the word 'degree', it is not strictly speaking a quantitative matter, but a qualitative one. However, Maynard has developed a six-point scale of participation which classifies the different ways in which people may be involved with the making of a decision. Under that scale, the six different degrees of participation are as follows:

(a) The workers make the decisions themselves (in a group, we should perhaps add: if each worker decides for himself individually it is a matter of delegation, not participation).
(b) The workers share in the making of the decision (obviously there are degrees of sharing, but we will leave these as subdivisions of this particular level).
(c) The workers raise issues for decision (this can be an important degree of participation when the principle of qualification is being used. The worker is frequently qualified to recognize and foresee problems, even if he is not particularly well qualified to produce solutions. It does not usually take a doctor to tell you you are ill, though it may take one to find a cure.)
(d) The workers are asked for opinions before the decision is taken.
(e) The workers are kept informed during the decision-making process.
(f) The workers are told about the decision after it has taken place.

These six degrees make a fairly clear-cut ladder of increasing participation. It is not so clear-cut, however, that moving up the ladder necessarily increases the extent to which the human objectives we have been concerned with are satisfied.

With group identification for instance, there is obviously a vast difference between being told nothing at all, and being at any step in the ladder. There is also a considerable improvement in moving from grade (f) to grade (e) and to grade (d). But thereafter the differences are not so great.

Anyone at grade (c), (b) or (a) is going to have his membership of the group emphasized. But a change in his authority within that group, which is what moving between them implies, does nothing further to make that sense of membership any stronger.

It might seem more probable that self-actualization necessarily increases as one moves up the ladder. But underlying that assumption is another one. It implies that the people being asked to make a contribution feel themselves competent to make it.

They may well be competent. A maxim of propagandists for participation is that employees at all levels have a much greater contribution to make than they are permitted to give. Where that is true, then fine.

But asking people to contribute at a level beyond that at which they feel competent does not increase their sense of their own value and importance (or even enrich their jobs). All it achieves is to subject them to a stress the effects of which may be as bad as those of the alienation they were supposed to be rescued from.

This point would match the principle of selection on qualification, but for one thing. That is that from the point of view of increasing someone's sense of value what counts is his own assessment of his capabilities. What counts from the point of view of improving decision-making is his objective capability. And the two do not necessarily coincide. There are plenty of people whose own view of their capabilities is higher than realistic. There are also a surprising number whose view of themselves is much lower than it should be.

As a problem, this mismatch of perceptions is not something the system designer himself can do much about. The only way it can be tackled is through adequate training programmes designed, not necessarily to increase skills or knowledge, but to heighten the participants' self-knowledge. The systems designer's task becomes one of producing a system that simultaneously encourages some people to participate who do not want to, and dissuades others who do.

Formality

Successively we have dealt with the scope, selectivity and degree of the participative system.

The last of the four factors that specify it we have called *formality*. Under this heading, the most important thing from the point of view of effectiveness and efficiency is the degree to which the system is formally instituted—that is, the degree to which it has set rules and formulae which must be followed. Again, it is not the case that formality is always good, or that formality is always bad. The difficulty is that increasing formality carries with it both advantages and disadvantages. The balance will have to be struck differently in different situations.

On the credit side there are several points. For one, a highly formal participative system helps to ensure a certain uniformity of behaviour throughout the organization. It helps to make sure that the participation sought is actually happening. For another, to satisfy some of the self-actualization needs it is necessary that participation should not only happen, but should be seen to happen. Participating may help to increase an individual's sense of his value to the organization, but unless it is seen to happen it does nothing to enhance his sense of status.

Moreover, for some people, making the system formal helps them to participate. People who are reluctant to put themselves forward in an informal situation may lose their shyness or timidity when their opinions are being formally sought.

In particular, in an environment which is soaked in traditional hierarchical attitudes, making participation formal may be the only way of making it work. This is equally true where etiquette and convention are dominant factors in the general social structure outside the enterprise: formal people need formal systems.

On the debit side, however, there are various factors of importance. Above all, virtually any formal system is bound to be slower than an informal one. It is also an invitation (which may be avoidable but is always present) to the creation of bureaucracy—what Rolf Lindholm of the Swedish Employers' Confederation has called 'gigantic apparatuses for the creation of dissatisfaction'. According to the same observer, it is quite feasible for a company with a highly structured participative system to find that 30% of its man-power-time is being spent on organizing and running consultation.

Moreover, the formal system is, virtually by definition, more rigid. One major benefit of the informal system is that it can, on occasion and when necessary, be dispensed with.

Partly for that reason, keeping consultation informal helps to avoid one major danger of participation in general. This is bound up with the fact that, at the extreme, the person rejected by a participative group is likely to feel even more strongly alienated than the person frustrated by a traditional hierarchical system. To have one's ideas blocked, or one's advice rejected, by an individual superior is one thing. It is at least easily rationalizable by assuming the stupidity, the bias or the arrogance of the superior. It leaves the individual on a common footing with the rest of his work group and may even increase his sense of identification with them.

It is a different matter when the ideas are blocked or the advice rejected by a group of colleagues. The rejection is that much more personal and therefore stronger.

It would be nice, of course, if people who found themselves at odds with a group were more likely to admit they were wrong than those in dispute with a single superior. But people's attachment to their own ideas is not all that strongly affected by the number of people opposed to them.

This is in spite of the evidence from several well-publicized laboratory studies. In those, a variety of stimuli (in one instance a number of beats on a drum) were presented to a group of 'subjects', all but one of whom were in fact disguised experimenters. Asked to say how many beats of the drum had been played, the disguised experimenters all give the same, incorrect answer. At least after a number of instances, the true subject ultimately suppresses his own, correct, view in favour of the majority, incorrect, one.

But this has rarely been evidence of an actual change of heart. Outwardly it may seem that the single subject has rejected his own views in favour of the majority. What in fact has normally happened is that he has remained convinced of his own rightness, but been too polite, or too timid, to express it.

Group persuasion is highly effective on outward behaviour, but not so effective on producing inner agreement.

Something of all this also underlies the paradox that the chief resistance to movements for participation, especially formal participation, comes from the least creative and the most creative sections of the workforce. The least creative prefer to accept other people's ideas and decisions anyway. And the most creative fear the inertia and conservatism of the group much more than the wilfulness of any single individual. Informal systems give them more chance of avoiding both.

It is easy to get trapped. In Yugoslavia, professional managers, in their frustration with the formally installed bureaucratic systems, have been known to commit sabotage: an ironic comment on the fact that sabotage by workers resulting from their frustration is one of the dangers participation is supposed to prevent.

There is therefore a difficult balance to be struck between two ways of achieving similar effects. There has to be a compromise between the two extremes: between setting up carefully structured and institutionalized mechanisms on the one hand, and on the other simply encouraging managers and supervisors to involve others in making the decisions they themselves still remain formally responsible for.

The major determinant of where the final balance should be struck must ultimately be the values and attitudes, the educational standards and social patterns of the people involved and the community to which they belong. It cannot be the application of any universal principle.

The parallel hierarchy

Before leaving the question of formality, we ought to make the point that formal and informal systems are to some extent different evolutionary stages, rather than opposites. Informal systems tend to evolve into formal ones. Formal systems can, of course, come into existence full-blown, especially through legislation. But more often they are preceded by informal developments.

It can happen in many ways. But in Fig. 8.1 we summarize a pattern of events that has taken place more than once. What has happened in this instance is that originally managers began to develop consultative processes within the hierarchy in order to counteract some of its weaknesses (step 2 in the diagram). Then, to cope with transient problems, some temporary consultative groups were set up outside it (steps 3 and 4).

Most of these were disbanded as their tasks ended. But some turned out to be worth keeping on a permanent footing (step 5). Then they began to develop links with one another (step 5), until the end product became

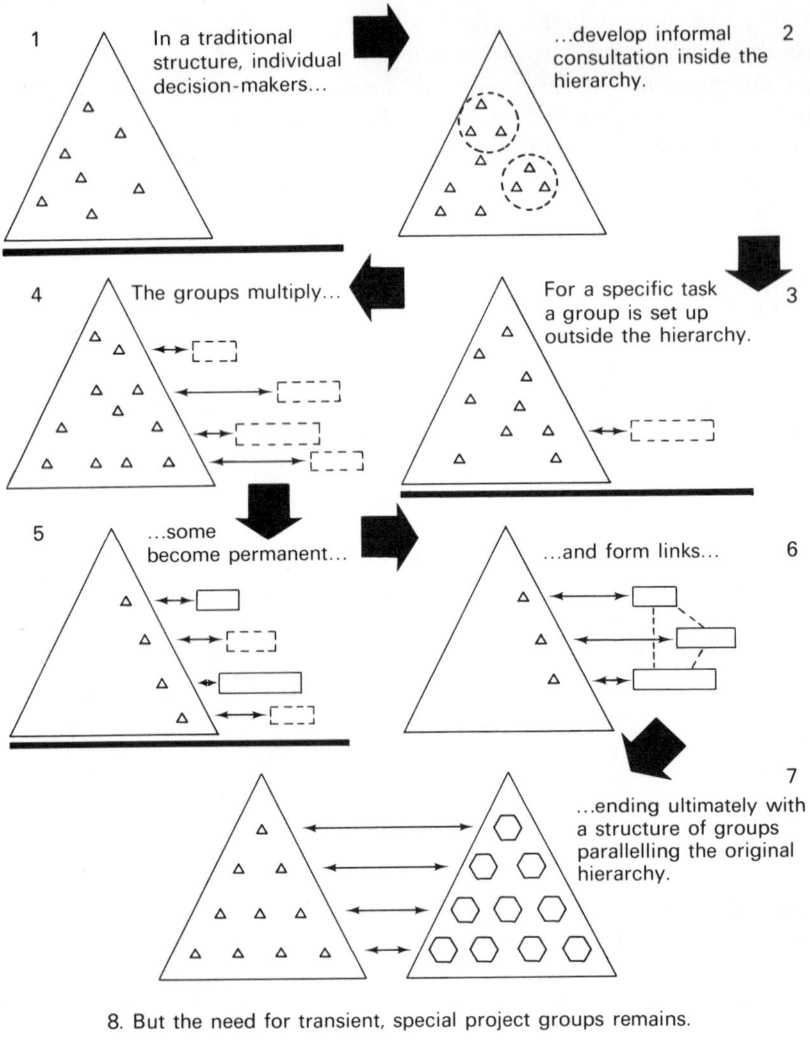

Fig. 8.1. Evolution of the participative hierarchy.

virtually a full-blown hierarchy of participative groups, parallelling the original hierarchy.

It is tempting to call such structures 'Swedish', and in fact the example given in Fig. 8.2 does come from Sweden. In it, the works committee is in effect a sub-committee of the works council. 'Reporting to it,' in the terminology of classical organization theory, are some line groups and some staff groups.

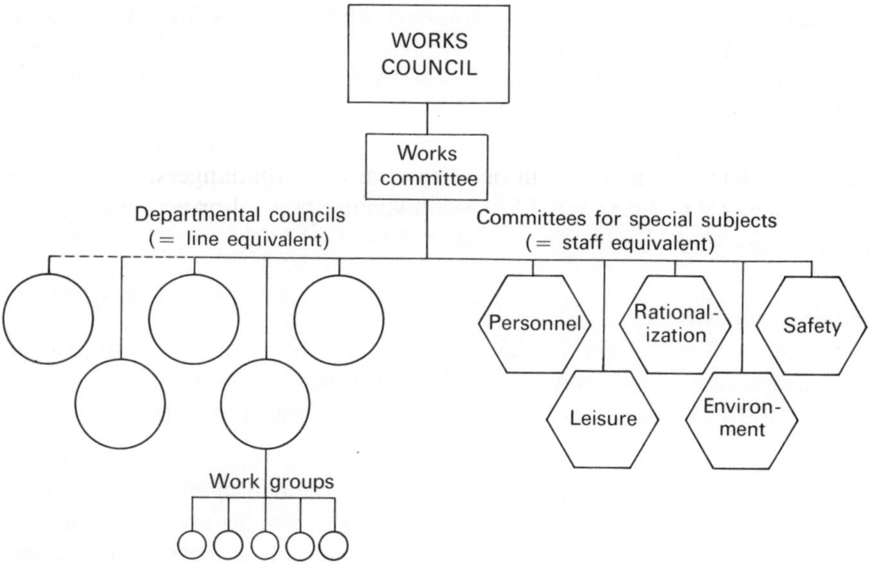

Fig. 8.2. The 'Swedish' model.

On the 'line' side are councils, with management appointees and elected workers' representatives, representing the major departments of the company. Each of these has direct links to the fundamental work groups in the department—each shop in the production department, for instance.

On the 'staff' side are committees consisting of works council members with special interests in specific policy issues.

The overall effect is indisputably that of a parallel hierarchy (though in this particular instance the 'staff' committees did not exactly match the responsibilities of the staff departments).

It is fairly easy to criticize such structures for their tendency to bureaucratization. But the 'Swedish' model is a not unreasonable starting point for designing any formal system. Its chief merit is that it allows decisions to be delegated to the proper level in the same way that the traditional hierarchy does. It therefore helps to ensure that those participating in any particular decision are both involved in the issue and knowledgeable about it. By allowing formal union representation (if desired) at appropriate levels, it also provides for participation on the basis of power.

All this may be saying no more than that in running a participative hierarchy one has to pay attention to the same factors that one takes into account in running a traditional hierarchy—in particular, the need to delegate. As long as those words of warning are borne in mind, the classic model will normally serve as a starting-off point, once some degree of formalization has been decided upon.

Summing up, therefore, we have so far described four factors in the design of participative systems that need to be taken into account: their scope, their selectivity, their formality, and the degree of involvement they allow. We have discussed the threats and opportunities, the promises and the dangers, inherent in each. And we have avoided laying down many universal principles.

Some warnings

A few general points are worth bearing in mind, however, even if in particular circumstances they may have to be disregarded.

From what we have discussed so far, we can draw out five points that it is essential to remember.

(a) It is important to balance the degree of participation with the desire for it that exists. One of the weaknesses of legislation is that it overrides this point. In meeting legal requirements, it may well happen that a system is installed that the great majority of members of the company are simply not interested in. And that is not just a futile exercise: it is worse. It almost certainly leads to excessive influence falling into the hands of the minority that does want to use the system.

In the absence of legislation, to go ahead and build a system on the untested assumption that people want it is an invitation to disaster. (So too, of course, is to refuse to build a system on the untested assumption that they do *not* want it.) Accurate and objective assessment of local feeling is critical for success.

(b) It is vital that all parties with power in the situation are participating in it. For efficiency, 'realpolitik', pragmatic recognition of where true power lies, is essential. Masking the realities out of a desire to keep a sense of status or propriety is likely to be fatal.

(c) Less emphatically, the degree to which participation is likely to be useful is inversely related to the specialized nature of the subject. It is not only for the traditional reason that 'the experts know best'. It is also because inviting people to participate in affairs in which they feel unqualified can create stress situations which are worse than the original problem.

One major thing that makes people feel unqualified to participate is newness and lack of experience. High labour turnover is therefore a factor that militates against successful participation.

(d) With some relevance to the last point, participation works best when people are used to it. With any system, there are likely to be teething troubles, which can only be lessened by intelligent anticipation and training. And since the best method of learning is doing, the advice that results must be: start small.

(We pointed out in chapter 1 that modern methods in education are increasing the degree to which young people are used to participation when they come into industry. As well as increasing the demand for participation, this development may also, therefore, increase the ease with which it can be introduced.)

(e) Unless care is taken, it is extremely easy to build up top-heavy consultative structures. Literally, it is possible to create thousands of jobs in an organization of tens of thousands of people. Not many organizations are in a position to carry that kind of supercargo.

Apart from inefficiency, such systems may result in an unacceptable level of centralization. And the people involved will resent the waste of their time if results do not justify it.

The need for common interests

Those five points summarize what we have said so far. But there is one final general principle with which we ought to close the chapter. This is probably more important than any of the previous points. It is that no kind of participative system is likely to be effective unless there already exists an awareness among the parties concerned that they share at least some goals. With it, participation can be fruitful. Without it, participation will be futile.

Of course, some of the essential interests of the various parties are different. Even managers' interests do not correspond 100% to those of their shareholders. And there has been plenty of evidence in recent years of the potential divergence of interest between a community of workers and their formal union organization.

But, while some of the fundamental goals may be different, they have to overlap. Or, put differently, there has to be some common means for achieving those different goals. There has to be common interest at the tactical and strategic levels.

Whether such common interest always exists, but is sometimes unrecognized, or whether it never exists, but is sometimes pretended, is not a point we are trying to decide. Our point is that, unless it is *believed* to exist, there is not a great deal of hope that participation will be successful.

Advocates of participation frequently claim that it can of itself *create* that sense of common interest. But that is an argument of which it is wise to be wary. There is undeniable truth in it. But it only applies where the shaping of the goals of the enterprise is itself participative.

That is, it only happens when those involved know that they are contributing, not only to the way in which operations are carried on, but also to the ends they are working towards.

Sometimes the argument is merely that 'allowing people to participate' in decisions will automatically make them readier to accept *your* purpose as theirs. And that is much more suspect.

From this point of view, in any organization there are likely to be three groups of people:

A: Those who already feel a sense of identity with the enterprise, and are aware that its interests and theirs overlap considerably. Some members of this group (we can call them A1) will already be participating in decision-making, either because of their managerial position, or through the union, or because they have some special qualities (it may be long service) that make their opinion valued.

Others (A2), however, are prevented from such participation for one reason or another. They may or may not feel frustrated as a result.

B: Those who are indifferent to the aims of the enterprise. It may be through apathy, or it may be through a positive belief that the enterprise's goals are irrelevant to them personally. Again, some may be participating already (B1) and some may not (B2).

C: Those who believe that the interests of the enterprise and their own are inherently opposed. Many of these are, in fact, likely to be already participating, if only through the union structure, or through unofficial action. Again, therefore, we have sub-groups C1 and C2.

The introduction of participation will affect each of these groups in different ways.

Group A, who already have a sense of common interest, will probably be positively affected. A2, who are gaining influence, will be the more strongly affected: A1, who already had it, less so. (Indeed, some members of A1 may be negatively affected if they believe participation is against the interests of the company, and therefore their own.)

But, as a matter of definition, it cannot be claimed that the sense of common interest has been created in group A.

Group C, those who believe in an essential conflict, are unlikely to be persuaded to change their mind by any management action. For them, the introduction of participation is more likely to be seen as a hypocritical trick.

It does, however, have two effects on them, one favourable, one unfavourable. On the one hand, it removes their ability to claim that they speak for all the workers (at least where the claim is not true). But on the other, it turns all the non-participating C2s into participating C1s, which is likely to be troublesome.

Group B, the indifferent, is the only group in which some sense of common interest may be created. As far as the B2s, the ones who do not already participate, are concerned, the novelty and flattery of being asked to participate may stir some out of their apathy. More, through participation, those who felt that the company's goals were irrelevant may learn otherwise.

(But the request for participation must be genuinely felt. Being asked for an opinion or advice because it is wanted is one thing. Being asked purely because some formal system says you have to be asked is another thing altogether.)

But the effect of the change on the members of B1, those who were already participating, will be different. For general participation threatens to reduce their own privileged position. By definition, they are unconcerned with the company's goals, in spite of the fact that they already participate. The argument that participation will help the company therefore does not move them.

Depending on the value that they place on their privilege, they will either be unaffected, or negatively affected. They may even turn into C1s.

Everything therefore hangs on how the workforce involved is split between the various groups. There may be techniques of education and persuasion (other than participation) that can help to move people from one group to another, but even these cannot be properly applied unless the initial position is known and understood. What is evident is that the simple introduction of participation alone cannot do much to change things.

We can therefore only conclude by repeating the basic message: participation can only be successfully introduced into any organization when the parties involved are aware that their interests overlap.

Example. Steelworks

When, in 1957, the Swedish government nationalized the Lapland iron ore fields belonging to the Gränges metals company, the money it paid in compensation provided the foundation for a successful programme of growth through diversification. By 1973, the group's sales had been boosted to nearly a billion dollars, and the company was operating in both ferrous and non-ferrous metals, heavy engineering, glass and shipbuilding.

During the process of growth, it encountered much the same problems of communication and control as any other large and growing corporation. Initially, it attempted to cope with these through grouping the independent operating units in five divisions. By the early 'seventies, however, the company's top management decided that this imposed an unnecessary third level of decision-making. The fact that there were group headquarters, divisional headquarters, and operational units, all concerned in decision-making, led to a great deal of bureaucracy.

In 1973/4, therefore, the five divisions were eliminated, removing one of the divisional levels. Instead, each of sixteen profit centres now operates with virtually full responsibility for its current business operations, and is responsible directly to the main group board. Only a few areas, such as financial planning, long-term corporate planning and personnel policy, are still managed from the centre.

So far, this is an exercise in classical organization theory. What makes Gränges different from most other companies that have gone through a similar evolution is the extent to which the reorganization has been accompanied by the introduction of formal participative mechanisms into the management of the company.

There already were worker representatives of the workforce on the main board of the group, but during the reorganization a great deal of emphasis was placed on increasing participation at lower levels—where the company is convinced the chief benefits from participation lie. In the view of the president, Johan Åkerman, 'the higher up you get, the harder it is for the workers to get any benefit from participation. At board level the only thing they are specialists in is the opinions of the employees. They can contribute little on how much we should spend on investments, for example. But it doesn't do any harm for them to listen to this. It helps them see that running a company is a very complicated business.'

In the new organization, therefore, participation is provided for at all levels. As an example, we can take the structure provided in the company's steelworks at Oxelösund, south of Stockholm, one of the company's 16 profit centres. In common with the other profit centres, the board responsible for running the operation itself has two worker representatives, one from the staff and one from the shopfloor. But, in addition, it has over 30 committees through which the plant's 3100 employees are more directly involved in decision-making.

At the top of these is the works council, with 12 members. Two of these are senior managers, two shopfloor workers, one a staff employee, one a supervisor, one a safety specialist and one a worker safety representative. The other four places, two for management and two for workers, are reserved for specialists in whatever happens to be the topic under discussion.

In addition, some of the regular works council meetings are open for any employee at the plant to attend, and also to put questions.

Reporting to the works council, and responsible for drawing up the agenda for its meetings is a much smaller steering group with a coordinator. He is also responsible for supervising the work of four sub-committees responsible for such matters as personnel and safety.

At department level, much the same structure is duplicated, with a department council corresponding to the works council, and shopfloor work groups corresponding to the sub-committees. Each plant sub-committee and each department council has a budget which it is free to spend as it sees fit on particular areas. Thus, in the safety area, the various groups involved had a total budget of $440 000 to spend in 1973.

At the lowest level, the supervisor meets with his work group once a month, usually when machinery is undergoing regular maintenance, but always during working hours. They are told how current performance is matching up to targets, informed by the supervisor of any problems he is aware of, and asked both for suggested solutions and for potential problems he may not be aware of. Workers are also free to bring up any matter they like concerning their jobs.

Many problems, of course, cannot wait for a monthly schedule before they need to be solved, and supervisors then need to make decisions the same way that they would in any other factory. But the work group always has the chance to discuss them afterwards, if they feel they need to.

Through the system, workers have an opportunity to play a direct part in such things as drawing up budgets for their departments, planning production up to a year ahead, forecasting costs, and recruiting new employees. Frequently, such activities need cross-reference to other departmental groups. Thus, a decision by one group of workers to change one workplace layout had to be referred to four other groups using the same equipment on different shifts before it could be finally resolved.

As of now, Gränges is fully convinced that the introduction of the system has improved employee attitudes and motivation. Oxelösunds has the lowest labour turnover of any steel company in Sweden, and, the company claims, probably the lowest sickness rate for any firm in the country.

Fig. 8.3. Examples of legislation for participation.

Country	1. *Size of firm affected* 2. *No. of worker representatives* 3. *Election of representatives* 4. *Nomination of candidates* 5. *Term served*	*Main council's rights and responsibilities* *(Subsidiary council's/committee's rights and responsibilities are given in parentheses.)*
Austria	1. More than 20 employees 2. Sliding scale: max 14 +1 per 500 employees 3. By all employees: in larger firms white and blue collar elect separate councils, which may sit together 4. By groups of employees 5. Three years	— agrees to: establishment of fixed salaries and piecework rates; working rules; transfer of salaried workers, if it involves a reduction in salary; holiday rotas; disciplinary measures; dismissal of members of the council — controls the application of legislation affecting working conditions, — participates in all inspections of installations — must be informed of all hirings and dismissals — at its request, must be informed monthly on the economic and financial state of the enterprise, and be allowed to participate in company planning — may conclude binding agreements with the management — may make representations to the appropriate authorities if it believes action is being taken contrary to the public interest — elects two members to the supervisory board of the company
Belgium	1. More than 50 employees 2. From 4 to 18, according to size of company 3. By all workers 4. Only by representative workers' organizations 5. Four years	— advises on, makes suggestions for, and raises objections to, measures affecting work organization, and working conditions — must be kept informed of the financial and economic state of the company — determines criteria to be applied for dismissals — establishes internal regulations, in particular on holidays and rotation of personnel — manages the social services of the company (Safety questions are dealt with by a separate committee for safety and hygiene)
Finland	1. Employing more than 120 000 man-hours per annum 2. Five or seven according to size of company 3. Secret ballot of workers	— must be given quarterly information on the state of the company including reports on technical progress — deals with matters affecting safety, hygiene and welfare — settles complaints — is otherwise purely consultative
France	1. More than 50 employees 2. 3–11 depending on size, more if agreed 3. By all workers: each union	— cooperates with management on all affairs concerning conditions of employment — manages all social services — must be consulted on all matters affecting the organization and the general development of the company

		— must be informed in advance of all dismissals — must agree to the dismissal of any of its members — appoints representatives to the safety committee (Workers elect separately representatives to deal with complaints)
Germany	1. More than five employees 2. Up to 35, if more than 9000 employees 3. Direct secret ballots of all workers 4. By 10% of work force, or 100 workers 5. Two years	— supervises applications of all legislation and collective agreements — examines, together with the employer, all complaints by employees — must agree: changes to working hours; disciplinary measures; rates of pay and changes to pay systems — shares decisions concerning the setting-up and running of social services — must be consulted on matters of security and safety — must be informed in advance of all hirings, transfers and dismissals, individual or collective — elects two representatives to the supervisory board (in companies with over 100 employees, an economic committee must also be formed with not more than four worker representatives, nominated by the council, and the same number of management representatives. It — must be given each month information on the economic and financial state of the company, including production and sales figures — reports every three months to the personnel of the company)
Holland	1. More than 25 workers 2. Max. 25 3. By all workers 4. Only by trades unions 5. Two years	— must be informed regularly of the economic state of the company — must be consulted on changes to working regulations — shares in the management of social services — may make suggestions for improving the economic and technical functioning of the company — otherwise has only a consultative role
Spain	1. More than 50 workers 2. 4–12 depending on size 3. No separate elections: those union representatives ('enlaces sindicales') gaining most votes in union elections are automatically appointed	— transmits suggestions and opinions from the workforce to management — controls the application of labour legislation — makes proposals for investment in social services — must be periodically informed of general developments in the company's affairs: production, sales, orders, etc. — may conclude collective agreements with the management — must be consulted over dismissals, suspensions, or changes in work arrangements due to lack of work — examines complaints, and attempts to settle them

Note: Elsewhere in Western Europe, Italy and the Scandinavian countries, while without specific legislation, have national collective agreements which produce almost the same effect. The United Kingdom has neither, though legislation is in preparation.

9. Managing change (1)

Promises, promises . . . Reluctance and resistance. Generation 1: decide and order. Generation 2: decide and persuade. Generation 3: ?

To design a production system, no matter how perfect, is not enough. It has to be made to work. Except when a new plant is being designed from scratch, change has to take place as the new system replaces the old. That simple fact alone has serious implications for the human objectives we have been bearing in mind throughout our discussion.

It has economic consequences too. Introducing any new system poses technical problems. New machinery and tools pose running-in and tooth-cutting problems. Stringent quality control mechanisms have to be applied to unfamiliar products. And, at bottom, there is always the financial cost of new equipment. But most of these are at least soluble in principle, even if they are sometimes difficult. And those that cannot be eliminated can at least be costed with some certainty, and thus allowed for. Either the economic payoffs are worth the economic costs, or they are not.

On the human side, the problems go rather deeper.

Resistance to change

It is true that, if the system has been properly designed, it should be superior to the old in terms of identification and self-actualization. At least it should be no worse. On paper it ought also to be true that the new system promises at least as much security as the old.

But there is much potential trouble hidden in that word *promises*. Promises are not the stuff on which security is based: but new systems can hold out nothing more than promises. Every new system, every planned change, is therefore a threat to security. It cannot be otherwise, because the threat lies in the very fact of newness.

Obviously, the threat may be large or it may be small. It is not so obvious (though it is in line with what we have said before) that the importance of the threat will be overwhelmingly determined by the nature and attitudes of the people involved. For some people, stability and security are of little moment. For others, they are paramount. But in every practical situation, they count. In one way or another, the system designer who is interested in seeing his system work in practice must overcome this fundamental resistance to change.

It has to be emphasized that resistance is not only a matter of objecting to changes for the worse. Recent years are full of instances of designers producing systems intended to offer the workers greater benefits in terms of all the objectives discussed here (more participation, better working conditions, more interesting jobs, and so on) only to have them rejected by the workers themselves. A bird in the hand is worth two in the bush. Benefits people already have outweigh much greater benefits that are still uncertain.

Whether the resistance is rational or not, however, it is not enough just to overcome it. It is not enough just to get passive acceptance of a new system. What is needed is an active commitment to it. This is a harder task, but it demands essentially the same techniques, and those techniques are the central subject matter of this chapter and the next.

It is sometimes claimed that it was not always that way—that at one time resistance to change was less than it is now, and that therefore the designer had to pay less attention to it then than now. But that would not be quite accurate. At least, put as baldly as that, it would be misleading.

Something as basic as the need for security is not likely to have changed much in only a century or two. Indeed, none of the changes we mentioned in chapter 1 have had any direct effect on resistance to change. It is not any increase or diminution in reluctance to alter work patterns that has lent new emphasis to the whole process of managing change in recent years, but a combination of other factors.

This chapter is devoted to examining that combination of factors, and describing the development of change management over the last two or three decades. The present situation is postponed to the next chapter. Anyone totally uninterested in history could perhaps pass over this chapter. But, in our view, an understanding of the way things have developed in recent times is critical to an understanding of the best way of tackling the problem in the present.

Looking, therefore, at that combination of factors that have given change management increased importance, we can put first among them the increasing awareness, on the part of managements anyway, of the *need* to change if the enterprise is to improve its results, grow and maintain growth or even simply to stay competitive and viable. 'To improve you must change. But not all change is an improvement.'

And the more often change occurs, and the more far-reaching it is when it does, the more prominent its associated problems become.

Moreover, while the changes of chapter 1 may not have made much difference to the worker's attitude to change as such, they have made a considerable difference to his ability to oppose it. More importantly even, they have increased his awareness of that ability, and his willingness to use it.

Once upon a time (the phrase is apt enough, since the situation may seem fairy-tale-like to many present-day managers) companies seeking to introduce change had several potent advantages. These made it possible for them to override, or even to simply ignore, resistance to change on the part of the workforce. More accurately, they could afford to ignore the worker's *reluctance* to change. For the reluctance was rarely translated into active resistance.

(a) Of these advantages, the primary one was *fear of unemployment*. Whatever the insecurity that is inherent in job changes, in the days before the welfare state it was far less than the insecurity involved in not having a job at all. (Among managers, as we have pointed out before, that is still often true. And managers are still less likely to translate reluctance into resistance.)
(b) There was also the *education gap* we have already referred to. One may not be happy about a particular change, but one is more likely to accept it if it is proposed by someone whose general education and knowledge is evidently superior. If one feels one knows just as much about the situation as the person proposing the change, one is more likely to oppose it. And that is more and more often the case.
(c) Finally, reluctance is more likely to become resistance when the mechanisms for *organizing* that resistance exist. In the context we are concerned with, it is not so much the increase in formal unionization that counts. Of more practical importance generally is the increase in the power of the unofficial and informal forms of worker organization at the shopfloor.

Generations 1 and 2

Once upon a time, then, management wishing to improve its production system only had to have the system designed by experts, and order its introduction. Then they could expect it to achieve results. All the manager needed to rely on was a desk with a loud, built-in, bang. That we can call *Generation 1* of change management.

It was not all that satisfactory even then, of course. Because what is needed for full effectiveness is not just passive acceptance, but active commitment, as we already noted. The difference was realized in what we can christen *Generation 2* of change management.

Generation 2, perhaps coincidentally, perhaps because of overlapping influences, was approximately contemporaneous with the spread of the 'human relations' school of management we described in chapter 2. Its attitudes are well enough exemplified by a passage from a book called *Job Economy*, which appeared at the height of the Second World War in Denmark. It provided Danish readers with their first opportunity of exploring what it called 'time-study and job simplification techniques'. Under the heading 'on encouraging cooperation from workers' it said:

> It has previously been pointed out that it is not normally necessary to aim directly at obtaining the operator's cooperation if only the interest of the production manager and the works manager has been won.
>
> During the study itself however the analyst must endeavour to win the personal cooperation of the worker, and this is attained by, among other means, discussing the job with him and as clearly as possible explaining to him the purpose of the analysis and the method to be followed.
>
> It must furthermore be made clear to him that his cooperation is of vital importance for obtaining a favourable result, and that this will be of advantage to the business and to himself in the way of easier and less tiring work arrangements.

It is easy enough to pick out from that the essential elements of Generation 2. The efficiency expert's first task was to get management approval. Then (though it was not 'normally necessary') he should attempt to win over the worker. He should explain to him the purpose of the change and the benefits that the firm and the worker himself could expect from it. Then, presumably, everybody would accept it.

Briefly, the principle of 'decide and order' had been replaced by 'decide and persuade'.

Generations 1 and 2 both had their successes, of course. Indeed, they were extremely successful. For this was the heyday of scientific management, and these were the methods used in producing most of the triumphant results we described in chapter 2. And we ought to try and account for those successes before we look at why and how Generation 2 finally ran out of steam.

One major factor was that, in a relatively unsophisticated industrial environment, it really was fairly easy to make simple changes that produced major results. The changes may have been small or big, but on the whole they were easy to understand. New ideas may even then have been difficult to discover, but once found it was pretty easy to see how the benefits would follow. They were therefore a more potent antidote to insecurity.

Reinforcing this factor was the fact that this was also the heyday of introducing piecework systems. The short-term benefits of increasing productivity were therefore brought more sharply home to the worker.

A more intangible factor, that it would be easy to overlook, was the missionary zeal of the experts themselves. They were protagonists of a new discipline that could be seen to be demonstrating its validity throughout the industrial world. Their own job, therefore, was of an almost ideal richness.

Intellectually, it was challenging. Looking back, some of the things that were done may look elementary, but at the time they required rigorous intellectual application. Their achievements were normally evident to others, and accepted by them. By and large, they were pretty free to do their job their own way. And their tasks were not only creative, but also varied and never trivial.

For these reasons, their own level of motivation was high. And the success of most missionaries is related to the fervour of their belief.

It was a belief that most managers and many rationally thinking unions came to share. Acceptance of the general principles of job rationalization time and motion study, industrial engineering, was widespread. It was not too difficult in the circumstances to persuade people of the merits of any particular application.

The same sort of factors helped the other group of experts, the practitioners of 'human relations'. And at that time at least there was no inherent clash between the two. The industrial engineer was concerned with the job itself. The human relations school, as we pointed out in chapter 2, was essentially concerned with the physical and psychological surroundings of the job.

Their spheres of action were therefore distinct. One was concerned with what Herzberg calls the *motivation* factors, and one with the *hygiene* factors. Each could afford to recognize the professional expertise of the other, and support it. In particular, the industrial engineer could usually rely on the support of the communication and information media developed by the human relations school.

Changing circumstances

Those were the favourable conditions that led to the success of Generation 2. Unfortunately, they did not last.

To begin with, as industry became more sophisticated, it became more difficult to produce the same scale of improvements from equally simple and inexpensive changes. Large benefits could be and were still achieved, but only at the expense of a greater degree of sophistication in the analysis, recommendations and investment required.

This was heightened by the fact that, because of their production successes, industrial engineers were asked to take in ever greater areas of the business. Production could no longer be treated in isolation from other segments of the business. Vertically or horizontally integrated rationalization offered greater potential.

As a result, not only did the demands on time and money made by rationalization studies increase, but the changes proposed became more and more difficult to understand. The benefits began to *look* less and less certain. The changes therefore became less and less acceptable.

At the same time, automatic respect for the 'expert' was diminishing.

Partly this was because of things we referred to earlier in the book: the lessening of the education gap, and a lower acceptance of authority and expertise in general. But it was undoubtedly helped by the change in the public image of the 'efficiency expert'.

When, in the 'forties, Hollywood finally produced a life of Gilbreth, the emphasis was heavily on the eccentricities and foibles of the pioneer's private life. In modern fiction, the efficiency expert has tended to be either villainous, or comic, or both.

Ironically, therefore, at the very time when the *true* education gap between expert and worker was beginning to increase (with the growing need for sophistication and rigour in the techniques of the specialty), the *perceived* education gap was lessening. The need for explanation went up just as the amount of explanation people were willing to put up with became less.

The increasing difficulty of producing the same scale of benefits had another effect, especially marked when companies were forced to introduce change simply to stay competitive, not in order to increase profits. This was that there was much less margin (if any) with which to pay financial rewards to workers.

Again, this happened just when their own increasing affluence was making it necessary to pay ever larger incentives to achieve the same ends. As we pointed out earlier, the more affluent the worker, the bigger the incentive has to be. It was therefore much more expensive to buy cooperation. The trend away from piecework increased the difficulty. With wages decoupled from output, the argument that better ways of working could produce direct financial benefits for the worker lost most of its force. The power of the economic incentive for accepting necessary change was lost.

A counter-argument was frequently advanced by opponents of incentives. According to that, decoupling wages from output would lessen fear of economic penalties resulting from change, and therefore facilitate its introduction. That argument, however, overlooks the need for a carrot in addition to security. If new methods are merely no worse than old ones, why change?

Finally, the motivation situation of the expert himself changed considerably. No longer was he automatically respected. No longer was he an innovator capable of bringing immense benefits. He was more likely to be seen as a conservationist clinging to old ideas that had lost their validity. While the job was as necessary as ever, it had lost much of its glamour. And, as the subject matured, much of its activity had become routine. Whereas one innovative section of industrial engineering has increased in refinement and expanded in scope sometimes, under the guise of 'management science', another section, under the guise of 'work study' and similar phrases, has become a simple basic function in the administration of many factories.

For people in that situation, the challenge of breaking new ground became less common. And as managements became better versed in the techniques themselves, the freedom from interference and the autonomy that the job had formerly had also diminished.

From this point of view, the arrival of those behavioural science experts that concentrated on the motivational aspects of the job itself did not help much. Whether they followed Herzbergian precepts or not, they broke the truce that had existed between scientific management and the human relations school. The spheres of action of the 'efficiency experts' and the 'motivation experts' now overlapped. And the net result was a diminishing in the status of the former, at least temporarily.

In the extreme, the efficiency expert, rightly or wrongly, tended to be seen as the traditional exponent of Theory X, and cast as a villain by those who took Theory Y as a commandment.

Such a reversal not only made listeners, both among managers and workers, less receptive to persuasion to accept change. It was also bound to have its repercussions on the experts themselves. They were on the defensive. And arguments launched from a defensive position are never as persuasive as those launched from a confident and assured belief in their validity.

The net result was that the 'decide and persuade' philosophy of Generation 2 began to lose its effectiveness.

It did not make the *need* for change any the less, of course. In fact, the need went on getting greater. What it did do was make it evident that a new philosophy for change management was required. It forced the development of Generation 3.

In the shaping of that new philosophy, several strands were involved. One of them was that there was an inherent contradiction between the new organizational systems that were now being introduced, and the 'decide and persuade' techniques for implementing them. Too often, Theory X was being used to introduce Theory Y.

If a production system being introduced paid a great deal of attention to giving workers opportunities for involvement and autonomy, it was illogical to present the system to them already worked out, and simply tell them how they were supposed to benefit from it. Exactly that was, and still is, frequently done. Not only was it more consistent to accept that, if the worker was to have an increased role in running the system, he ought also to have some role in its design. It was also logical to expect a higher degree of commitment to success if he were given that role.

Generation 3: ?

It had already been recognized in the days of Generation 2, of course, that it was desirable to talk *to* the worker before the decision was made. But this was essentially a matter of timing: better to tell him before than after. It was also essentially one-way: telling him about it, rather than listening to his ideas. Decision and persuasion remained separate things.

With Generation 3, the emphasis has become different. The decision and the persuasion fuse together in that the keynote becomes the drawing of the people concerned into the making of decisions themselves. The need for persuasion then fades away. The decision itself should incorporate the beliefs and views of those who otherwise would have had to be persuaded.

In the terms of the last chapter, the decision becomes participative, and the participation is at what we called level (d). The worker shares in the decision. Any lower level than that is an elaboration of Generation 2. And any higher level, except in occasional special situations, breaks the fundamental principle of authority going with budgetary responsibility. (The special situations are those where the workers also share budgetary responsibility. But these are rather rare, especially when what we are talking about is mainly the introduction of new production systems.)

Before we move on to the next chapter and a consideration of the practical application of Generation 3 change management principles, there is an apparent contradiction that needs explaining. We pointed out in the last chapter that the desirable level of participation in an organization would vary with the local situation and the issues involved. It may look like a reversal of opinion to state now that in the management of change participation needs always to be at a specific level.

But we are now dealing with one specific issue: the introduction of new or redesigned production systems. It happens to be an issue in which all three of the principles of participation come together. And, whatever the local situation, they all operate at much the same level. Whenever a new production system is being introduced, three things are always true. The worker has the power to affect directly the success of the innovation. He is obviously deeply involved in and affected by it. And his own skills and experience are likely to be useful and relevant.

We can therefore break our earlier rule, and be specific. And we can sum up the story of the three generations by saying that 'decide and order' and 'decide and persuade' have both given way to 'decide together'.

10. Managing change (2)

The consultant's roles. Six factors. Choosing projects. Managing projects. Preparing the ground. Organizing collaboration.

To say that Generation 3 of change management means 'deciding together' does not, of course, imply that management loses the final say. That is not conceivable if management is to be held finally accountable for success. It implies rather that management's final position should be found among solutions and proposals jointly recommended by several relevant interests and groups, including the workers concerned.

The consultant's roles

In any major change project, the process of arriving at that final position is virtually certain to be a long and complex one. The trend is for each project to require the involvement of more and more of the total enterprise. It becomes less and less possible or desirable to isolate the various functions from one another.

The question of participation by different levels in the hierarchy is therefore inextricably bound up with the question of interdepartmental, interfunctional and interdisciplinary participation. Moreover, many major projects are likely to require participation between permanent staff and temporary staff or outside consultants.

On the whole, however, when we use the words 'consultant' or 'expert' in this chapter, most of our comments will be equally valid whether the person concerned is an outside consultant or an internal expert. The critical point is his expertise, not his formal status. Whether he is being paid a fee for an assignment or a salary is not itself of any great importance in this connection.

That said, however, it is worth spending a little time considering the roles that the consultant/expert can play in the change management process. Partly, this is because it helps to see where the difference between outsider and insider might be important. But it is also one way of bringing together the various processes that have to be gone through in implementing a project, whether participatively or not.

There are various ways of classifying such roles, but basically there would seem to be seven ways in which the consultant can play his part as 'agent of change'.

(a) He can stimulate top management and employee interest.
(b) He can provide relevant experience and data from similar undertakings elsewhere, as well as special or proprietary approaches to solving particular problems.
(c) He can formulate the scope and purpose of the project and estimate potential results.
(d) He can programme the project, selecting optimal technical procedures.
(e) He can calculate the resources needed and plan their allocation, thereby estimating the cost and profitability of the project.
(f) He can train people in whatever skills are required (including the basic skills of project management).
(g) He can expedite the project and maintain its momentum.

Strictly, all of these could be performed by insiders or outsiders. Perhaps the hardest for an insider is (b), unless one envisages a continual recruitment of staff from other companies and industries, and allocating adequate time for the analysis and compilation of their information.

On the other hand, (c) is perhaps the hardest for an outsider. An outside specialist may be invaluable in suggesting possible objectives for a project, and in helping to clarify them. But determining what objectives *should* be aimed at (long-term anyway) is essentially an internal function. Any major project ought to tie in to some strategic plan. And determining the strategic aims of a company has got to be the responsibility of the people running it.

The other roles are ones in which both outsiders and insiders can play valuable parts. Whether outside help is necessary will depend, obviously, on what level and amount of expertise is available. Frequently the best solution is likely to lie in collaboration between the two, just as normally there will need to be collaboration between different sections of the company. For role (g), expediting the project, an outside consultant's neutrality between such collaborating sections may well be helpful.

Of more importance to the general theme of this book, however, is the subject of collaboration, not between outsiders and insiders, or between different sections, but between different levels of the hierarchy. In general, that is, between workers and management.

Major factors for success

For that discussion, we need a framework. Many might be possible, but we would like to use the set of concepts illustrated in Fig. 10.1. This picks out as the six most important factors determining the success of any particular exercise in change management:

— quality of project selection and initiation
— industrial engineering competence

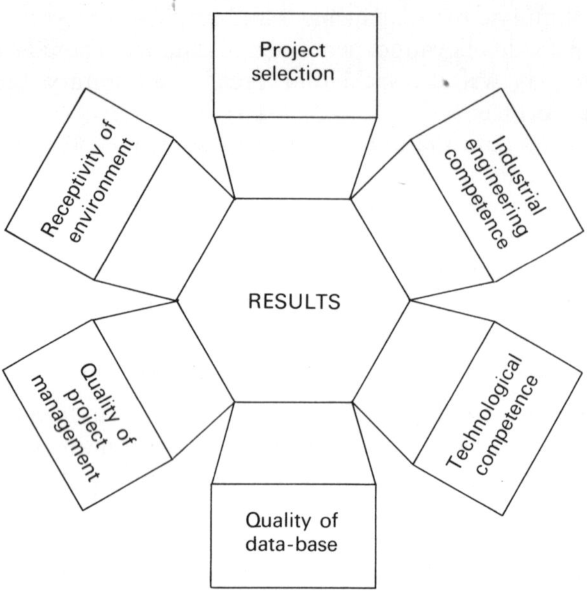

Fig. 10.1. Factors determining the results of change projects.

— technological competence
— quality of data base
— quality of project management
— receptivity of the environment.

Of these, three (industrial engineering competence, technological competence, and quality of data base) are essentially unaffected by changes that have taken place in the worker's role. It may be important to underline that all three are still necessary. Sometimes, in the rush to 'participation' as a panacea, it gets overlooked that professional skill in the development of the new technical system, technological competence in selecting and designing new processes and equipment, and the provision of a comprehensive and accurate data base on which technical decisions in all fields rely are still critical essentials to its success.

In some cases it may be that workers do command the necessary skills, but generally speaking these are areas in which the specialist role must remain the dominant one.

The other three factors all warrant longer discussion. We can take them in the order given.

Project selection and initiation

It may seem too obvious to assert that, if one chooses the wrong projects, one cannot hope for success. Nevertheless, we assert it again, since there is little point in paying a great deal of attention to doing the wrong thing right.

The selection process consists of two phases, which are separate though they may overlap in time. The first is the generation of possible projects. The second the choice between them. Running as a common thread through both of them is a third factor: determination of the strategic objectives the company is seeking to attain.

Much could be and has been said and written about the technical aspects of all three. We do not seek to make any general contribution to those aspects.

All we can do is try to give a brief view of the position, and emphasize those things which bring us back to the role of the worker.

From that point of view, one factor dominates the generation of objectives. We have mentioned it before. It is that, if people are to feel genuinely committed to the achievement of objectives, then it is virtually essential that they have some hand in setting them.

The question of worker participation in setting the goals (or some of them) for any particular project is not a matter of style but of effectiveness. For full effectiveness, it is not enough to ask the worker 'How do you think we should go about this?' (though that is better than nothing). What needs to be asked (genuinely, not formally) is more like 'What do you feel we should be trying to achieve?'

This is related to the point we made about the generation of objectives being the least likely area for reliance on an outside consultant. He may be able to stimulate and provoke. He may be able to suggest possibilities that might otherwise be overlooked, point to key figures and achievements elsewhere, or warn of impracticalities in some goals that are suggested. But essentially it is the people involved who need to set the objectives they are going to attain. And that means management and workers in some combination or another. It does not, usually, mean consultants.

In thinking up ways of attaining those objectives—that is, in the creation of ideas for projects—the consultant may well have a more significant role to play. On balance, he is more likely to have seen what has succeeded and what has failed elsewhere. And he is perhaps less likely to have his vision limited to improvements in the present situation, and therefore more likely to be able to envisage a completely new departure.

Which brings us to a basic point. Given a set of objectives, there are two ways of trying to create a project (or a set of projects) to attain them. One starts from the present situation and seeks ways of improving it. One starts from the objectives, attempts to set up an ideal solution, and then works back from that to a practical one.

It would be tempting to assert that the first way, the 'now-centred' way, is better applied to minor projects and the second way, the 'ideal-centred' way,

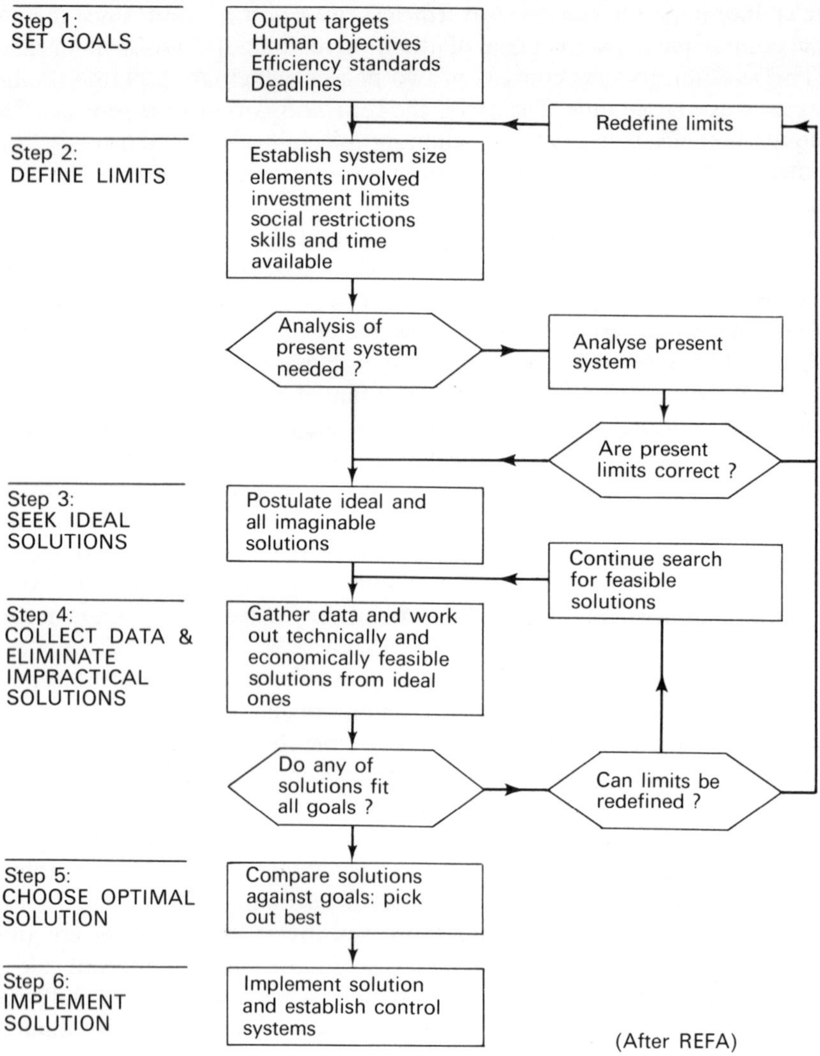

Fig. 10.2. Goal-directed design procedure.

to major projects. And there would be a lot of truth in the assertion. It would probably be even truer if 'minor' projects are defined as those that can be handled within the routine organization, whereas 'major' projects are those that require a special organization to be set up (of which more later).

But that masks the more important point. That is that, while the 'now-centred' approach is likely to be simpler, and lead to less disruptive changes, it also offers less chance of major gains. Conversely, the 'ideal-centred' approach, while it is likely to be more complex and lead to more far-reaching changes, also offers the potential for far greater benefits.

For our purposes, we do not need to dwell on the technical advantages of one method against the other (though we illustrate the second in Fig. 10.2, based on material from the German REFA-Institute). From the point of view of our major concern, the important distinction is that the balance of influence in the management–worker–consultant participation is desirably different in each approach.

The more closely anyone is involved in the present situation, the less likely he is to be able to divorce his thinking from it, but the more thoroughly he is likely to understand it. For minor changes, improvements to the present situation, projects that seek relatively slight returns, worker participation in generating ideas is likely to be most important.

Choosing among suggested projects is a different matter. In essence, it involves assessing each project against each of the objectives (the 'project criteria') and estimating which on balance does most to satisfy them as a whole—while staying within necessary constraints of money, time and other resources. Whatever the case with setting objectives, the business of estimating effects, of calculating how likely any project is to meet the objectives, is something that requires a much higher degree of specialist attention—and therefore provides less room for participation.

This is equally true whether the objectives are financial, technical or human. But all these varied tangible and intangible objectives have to be balanced together. And, in the balancing, participation becomes important again. Not, perhaps, worker participation alone, but certainly participation between different specialists, managers and workers.

For specialists, be they accountants, engineers or industrial psychologists, are only too likely to give too much weight to the objectives they personally are concerned with. And, in balancing their views, participation by non-specialists is indispensable.

Coping with objectives which cannot all be measured by the same yardstick, or may objectively not be measurable at all, is obviously fraught with problems. Indeed, except in the rare cases where one project is evidently better than all the others on every count, there may be *no* objectively logical and mathematical way of choosing the 'best' of several projects.

But that is nothing to be ashamed of. If mathematical rigour is impossible, it has to be dispensed with. If logically perfect answers are unobtainable, we have to be subjective. And where we have to be subjective, the feelings of those involved in the situation permanently are likely to be more valuable than those of the outsider, or even the internal specialist whose attention is only transiently given to the situation.

Being subjective, however, does not necessarily mean not being systematic. To be systematic in this area, any approach must include:

- giving a weight to each objective
- assessing the degree to which each project meets each objective
- combining the weights and the degrees into some kind of total score for each project.

There are many ways of tackling each step, of course. A practical one is to give each objective a rating (jointly arrived at) on a 1–10 scale, do the same for contributions, multiply the contributions by the relevant objective weights, and total all the products for each project. The sum then becomes the score for each project. (Part of a more complex method, given as it was used for one German company, is illustrated in Fig. 10.3.)

Of much greater importance, in practice, than how the assessment is done is who does it: consultant, specialist, manager, supervisors, workers. There is no universal answer, except that no one of those groups can do it effectively on its own.

That is no more than a sketchy description of the problems involved in project selection. But it may have served to point out some of the pitfalls, and perhaps some of the opportunities, that lie in participation.

Project management

We have used the term 'industrial engineering' to refer to the technical skills involved in designing and specifying the new system. By 'project management' we refer to something different: the ways and means whereby change is systematically organized and administered.

In determining the quality of project management, probably the key factor is understanding and controlling the complex relationships between the people involved: between permanent staff and temporary staff; between specialists and non-specialists; and between the various kinds of specialists themselves, as well as between management and workers.

Of course, the project manager has to have all the traditional management skills. Of course, too, he must be dedicated to the success of the project, which usually means he must have had a hand in creating the idea. And of course he must be supported by efficient administrative procedures for controlling results, time and costs. But these are things that apply to any manager, anywhere, in any job. What distinguishes the successful project manager is precisely his ability to cope with the complex web of interpersonal relationships we have mentioned.

That still leaves it difficult to answer the key question of which the manager should be: staff man or consultant, line man or specialist. It is probably true that a good project manager can emerge from any group: so too can bad

	OBJECTIVE													Total	Weight
	B	C	D	E	F	G	H	J	K	L	M	N	O		
A	A3	A2	D3	E2	F3	G2	H2	J2	K2	L1	A1	N3	O3	6	2
B		C1	D3	E3	F3	G2	H3	J3	K3	L2	B1	N3	O3	1	–
C			D2	E2	F3	G2	H2	J2	K3	C2	C1	N2	O2	4	2
D				D1	D2	D3	H1	J1	K1	D2	D3	D1	O2	20	7
E					F2	G1	H3	J3	K1	E2	E2	N1	O3	10	4
F						F1	H1	J1	F1	F2	F3	N1	F1	19	7
G							H2	J2	K1	G2	G2	N2	O1	11	4
H								H3	H2	H3	H3	H1	H1	27	10
J									J1	J3	J3	N1	O1	21	8
K										K3	K2	N1	O3	16	6
L											L1	N2	O3	4	2
M												N3	O3	–	–
N													O1	19	7
O														25	9

Each objective is compared with each other objective and the result entered in the corresponding cell as follows:

1 The letter of the more important objective.
2 An estimate on a 1 – 3 scale of the degree of priority of one objective over the other.

The scores for each objective are then totalled as shown by the arrow for D. The highest ranking (H) is given a weight of 10, and the others pro rata weights according to their totals.

In the case from which this example was taken, the letters stood for:

A Physical and mental demands on the worker
B Maintenance costs
C Qualification required from workers
D Liability to disruption
E Possibilities for development for workers
F Controllability
G Variation in cycle time
H Simplicity of material flow (in production)
J Simplicity of material flow (from stores)
K Motivation of work force
L System flexibility
M Area needed by layout
N Simplicity for personnel planning
O Productivity improvement

Fig. 10.3. Weighting incommensurate objectives.

ones. However, it is safe to assert that a good project manager is either a good executive, or a good executive candidate.

The outsider has the advantage of established neutrality. The insider has the advantage of established status and esteem. The outsider is likely to have more experience of similar jobs. The insider will know the people with whom he has to deal. The balance is evenly tilted. The logical conclusion, therefore, is that what is normally needed is collaboration.

Ways of organizing that collaboration we will come to in a moment. First we should deal with the last of the factors of change management presented in Fig. 10.1.

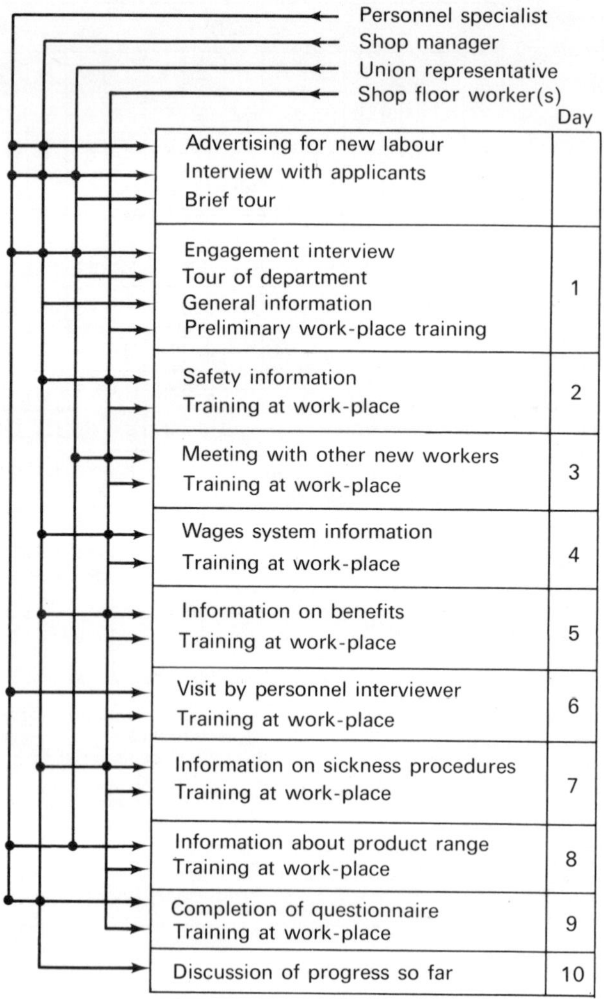

Fig. 10.4. Teamwork in the induction of new employees.

Receptivity of the environment

With this factor, we are back to the subject of the last chapter: people's attitudes to change. To what we said there we have only one thing to add now. That is that the 'decide together' philosophy of Generation 3 of change management will be normally more successful where participation is already a feature of daily life.

That may well mean that, with an eye on making future changes easier, more participation will have to be built in to the routine management of the enterprise than would otherwise be warranted by the considerations of chapter 8. It need not be anything dramatic or high-level. Figure 10.2, for instance, illustrates the programme followed by one Danish company in the induction of new employees. It is an effective way of introducing people to their jobs. It has also had a considerable effect on reducing labour wastage. Formerly the company had a large proportion of workers leaving after a few months. Now most get through that critical stage and settle in as long-term stable employees.

But, from our immediate point of view, it also accustoms workers, their representatives, local managers and personnel specialists to working together at something not part of their day-to-day jobs. It therefore fosters the kind of attitude to collaboration that pays off when a major change needs to be introduced.

A more fundamental approach to altering the receptivity of the environment derives from what is normally called *matrix organization.* A company with a matrix organization has, permanently, a kind of dual hierarchy. Thus, a functionally organized company may also have individuals who are responsible for certain products, or groups of products.

A salesman or an accountant, say, will therefore have two bosses. He will still be responsible to his functional superior for his speciality, but will also be responsible to the product manager for his work on that product.

Matrix organization is normally introduced as a more effective way of carrying on routine operations. But, under it, individuals, both workers and managers, become accustomed to dual relationships. Workers (and managers) become used to working with two superiors. Managers become used to having only the part-time allegiance of subordinates. It creates problems, of course: but the point is that, once the system is established, people become used to coping with those problems.

And, since the problems are similar, people working routinely in a matrix organization will find it easier to adapt to the demands of project management in the sense we are talking about.

Indeed, the two are so closely connected that the permanent stage of matrix organization is sometimes confusingly called 'project organization'.

One could dream up many ways of building an organization to cope with the problems we have been describing. Many or most of them would probably work, as long as the various requirements we have stated are kept in mind. One way that has stood the test of time is that illustrated in Fig. 10.5.

At the heart of this system is the steering committee and its chairman. Both chairman and members would normally be appointed by the senior

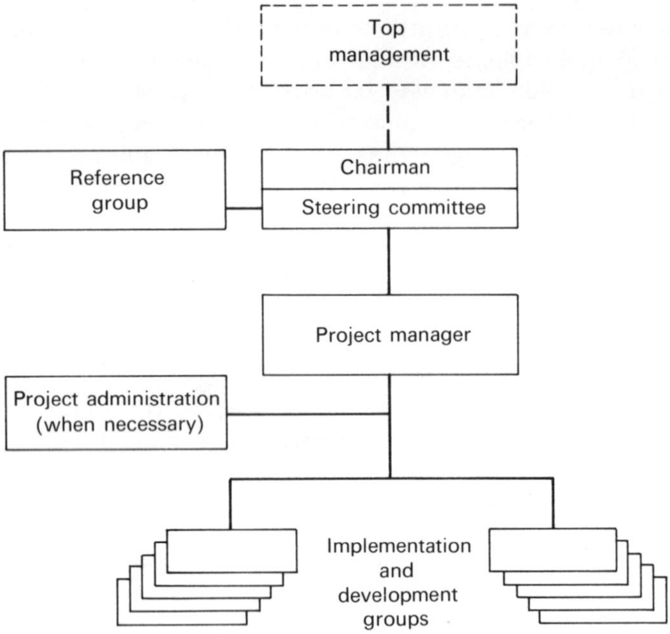

Fig. 10.5. Project organization.

management group (sometimes the board of the company); though where worker representatives are included, a union might be invited to nominate them.

The actual decision power usually rests, not with the steering committee as a body, though it is expected to operate participatively, but with its chairman in accordance with the responsibility given him by the board, or with his executive authority in the basic organization.

The steering committee which he leads will be a small group of people meeting periodically to follow up the progress of the programme, receiving written and verbal reports from the project manager (or, on occasion, project managers). Proposals for specific decisions are presented and responded to. The steering committee will normally represent the line and staff functions involved in a major way in the project. There may, however, be more peripheral areas of knowledge, experience, and interest which are still important to the project. There will, in that case, be persons whose opinions and advice will be sought by the steering committee, through what we have called in the diagram the 'reference group', though it might well bear a title like 'advisory committee'.

The same group can also be used for the simple transfer of information about the project.

In the next box down we have placed, not a group, but a man: the project manager. There will always be a need for someone charged with the full day-to-day control of the overall project and the elements that go to make up its programme. In complex situations he may well require an administrative team working directly for him.

Below the project manager we have indicated a number of different groups charged with the development and implementation of different aspects of the programme. Each of these would need to be headed by the project manager or an assistant group leader. Each might include workers or lower level managers as well as specialists (internal or external).

To some extent, the possible role of the consultant in the organization varies as one moves from top to bottom of the chart. At the top, he is more likely to be working as a stimulant, guide and troubleshooter. In the middle, he is likely to be operating as an advisor and technical specialist or as a project executive. At the bottom, his role is more likely to be as an expediter or trainer. His roles as an expert calculator and planner, and as a source of outside experience, are likely to be of more or less equal importance all through.

It is worth noting that the consultant in this type of organization may well have a useful role to play as project manager. We have already indicated the relative weaknesses and strengths in this situation of the insider and the outsider. But if an inside figure of adequate status is chosen as steering committee chairman, and an outside consultant as project manager, then close collaboration between the two can overcome the weaknesses and build on the strengths.

Figure 10.6 indicates the relationship that such an organization may have to the basic 'standing' organization of the company (the management hierarchy) and the representative organization, where there is one. At the risk of a little confusion, we have deliberately chosen to show the project organization in a slightly different form, and with slightly different titles. It seems worth making the point that variation is easily possible.

('Work groups', as may be obvious, are effectively sub-committees of the steering committee concerned with specific aspects. They may also include people who otherwise might be part of the reference group.)

The more important part of the diagram is at the bottom, showing how the proposal worked out by the steering committee and its work groups is submitted to the representative organization (at whatever happens to be the appropriate level) for review, and to the management hierarchy for decision. Installation then becomes the responsibility of the project organization, which also is required to keep both parts of the standing organization informed about what is happening, and educated to understand it.

This chart brings us to the end of this chapter and our discussion of pro-

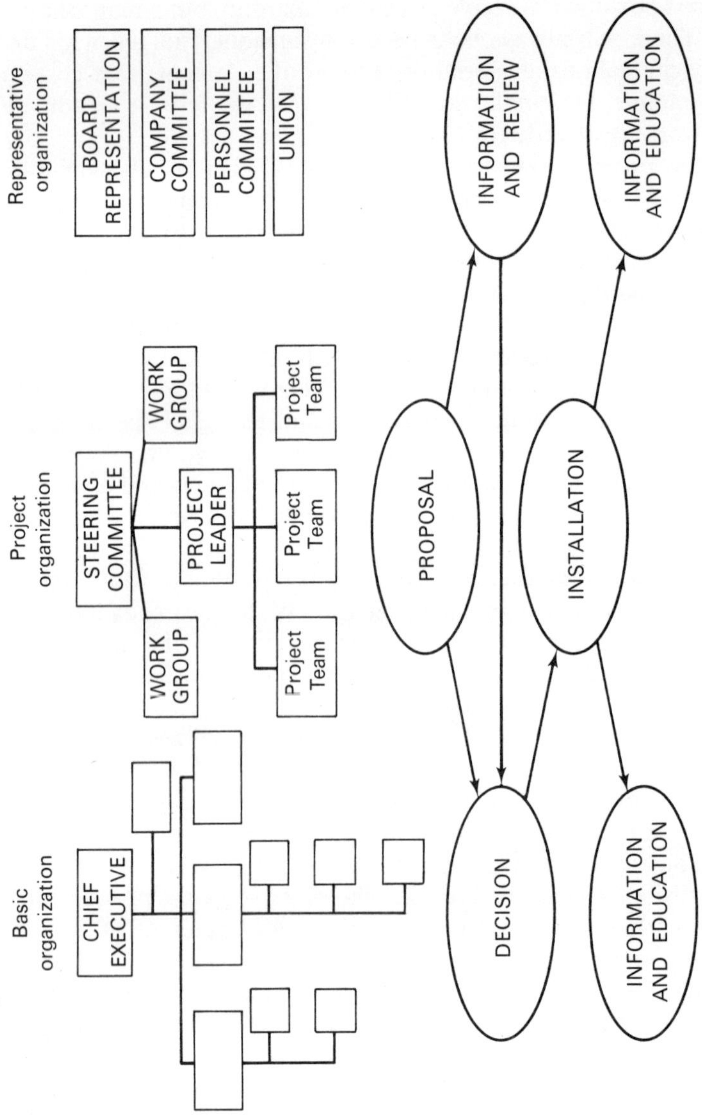

Fig. 10.6.

ject management. We have departed from the practice of the earlier part of the book in not including examples after this chapter and the last. However, the longer case studies which are included in the Appendix illustrate the points we have been making. They are all examples of major change projects.

11. Summary

The argument—and a conclusion.

1

After two hundred years, there are signs that the Industrial Revolution is running out of steam, at least in the developed countries of the West. The principle of unrestricted private enterprise has already been abandoned. Now it also seems that another pillar of traditional economic thought will have to be considerably modified.

This is that economic efficiency can always be increased by the finer and more accurate dividing up of tasks. The division of labour has for many years been opposed on aesthetic grounds. Destroying the meaning of a job by breaking it up into efficient pieces has almost always alienated the intellectual and the artist. But it is only in recent years that it seems to have become self-defeating even from a strictly economic point of view.

That change can be accounted for by the realization that five assumptions about human behaviour which were fundamental to the Industrial Revolution are no longer true. Or at least are no longer as valid as they once were.

These five assumptions are that:

(a) People will do as they are told by those in authority.
(b) People will accept that their superiors know better than they do.
(c) Man's prime motivation is economic gain.
(d) Man's physical security depends on his having a job; and his having a job depends on his taking instructions.
(e) Man's sole interest is the welfare of himself and his family.

The extent to which these assumptions are still true varies from country to country and from community to community. But over the western world as a whole each is subject to considerable doubt.

As a result, the extreme and singleminded exploitation of one of the traditional principles of industrial engineering—that jobs could be broken down into ever more detailed and circumscribed tasks—has, in many circumstances, become unprofitable.

Yet, for many reasons, the West still has to face a challenge to increase its productivity. It needs to produce more efficiently, even if it does not need to produce more. Otherwise it can neither meet the competition of the developing world, nor satisfy the demands of its own people for an improved quality of life.

Meeting that challenge will require finding ways to unite the 'old' economic values of the early Industrial Revolution with the 'new' human ones that have emerged into prominence.

2

In the past, attempts to find ways of increasing productivity have been dominated by the existence of two schools of thought. One is that which flows from the work of F. W. Taylor and others, which we call *scientific management*. The other, normally asserted to stem from the famous Hawthorne studies, we call *behavioural science*, though in its early manifestations the term *human relations* was more common.

Each has changed and developed over time. But the critical distinction between them is that scientific management has focused its attention on the job, while behavioural science has concentrated on the worker. In recent years, behavioural science has emphasized the importance of the work itself rather than the surrounding circumstances, but it has still concentrated on the way in which the work satisfies the human needs of the worker.

To concentrate solely on human needs or to concentrate solely on economic objectives is a one-eyed way of looking at the problem of production

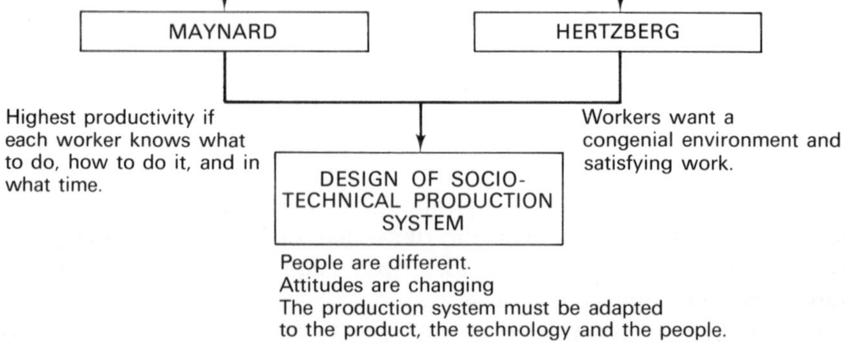

Fig. 11.1.

system design. If we are to meet the continuing challenge to increase productivity, we need a two-eyed stance. And that, as we illustrate schematically in Fig. 11.1, means taking into account both the work and the worker. It means considering simultaneously both economic and human objectives.

3

The process of production system design can be broken down into four phases: the survey phase, the resource phase, the system phase and the sub-system phase.

In the *survey* phase, the objectives of the system are determined. Its scope is also settled, as are the various resource limitations that may be placed on it.

There are two ways of looking at the *resource* phase. In the simpler way, it can be viewed as determining how much of the system should be dependent on hardware, and how much on software. From a more complex viewpoint, in this phase a balance needs to be struck between human resources, the capital resources and the system resources to be employed in the system. Within each resource area, other rough balances need to be struck. In the human resource area, this is between productivity and motivation objectives. In the capital resource area, it is between stocks (and work-in-progress) and the level of customer service. And, in the system resource area, it is between system power and system scope.

The third phase, the *system* phase, is concerned with the overall relationships between the various sub-systems that go to make up the whole production system. There are six of these sub-systems: the production mode, the job, the work planning and control system, the compensation system, the physical environment, and the organizational environment.

Each of these has potentially a different contribution to make to meeting the economic and human objectives the system will seek to fulfil. In the system phase it is necessary to consider all these impacts, and to determine what part each of the sub-systems shall play in achieving the system's overall goals.

(The potential impact of the various sub-systems on six selected objectives is illustrated in Fig. 3.2.)

Finally, in the *sub-system* phase, which will normally be carried out in different groups, each of the sub-systems has to be considered in detail.

4

In considering the *production mode*, the chief danger is making the assumption that there are only a few classical 'packages' to choose from. For example, one such package is the classic 'assembly line' picture, in which a mechanically paced line is accompanied by finely divided tasks, each performed by an individual on his own.

Another such package is the functionally organized machine shop for batch production purposes. In that, products are expected to be manufactured in relatively short runs, and machines are grouped according to the function they perform, and manned by craftsmen, each operating his own machine.

However, viewing the production mode problem as one of selecting the right package from a small number of packages means overlooking many possibilities.

A more fruitful approach is to recognize that there are three independent components that go to make up the production mode, and that a choice can be made on each component without affecting the others. These components are:

— *production flow* (the way in which the product moves through the plant during manufacture or assembly)
— *work orientation* (the way in which tasks are specified)
— *worker assignment* (whether workers are assigned to tasks as individuals or in groups).

Again, these different components affect the human and economic objectives of the production mode in different ways. Those objectives can be simplified into six. The economic ones are job simplicity, controllability, and flexibility. The human ones are product identification, group identification, and self-actualization.

How the different components affect the different objectives is shown schematically in Figs. 4.1–4.4.

The important general point, however, is that by recognizing the independence of the three components it is possible, for instance, to maintain the largely economic advantages of the traditional assembly line while avoiding many of its largely human disadvantages.

5

A similar confusion tends to haunt the subject of *work planning systems* since the choice is often made to appear to be one between centralization and decentralization. Even where there is seen to be a possibility of choosing different degrees of decentralization, the confusion still exists.

A more accurate position is that there are, again, three components of decentralization that are frequently mixed up. These are:

— *delegation* (the pushing of decision-making authority further down the organization)
— *participation* (the taking of decisions by groups rather than by individuals)
— *aggregation* (the principle of dealing, not with individual items, but with logical combinations of them).

Again, these have to be looked at against the range of objectives that the planning system designer has to keep in mind. Briefly, these can be stated as:

- *system economy*
- *system power*, which is broken down into
  - *diagnostic power*
  - *control power*, which is broken down into
    - *comprehensiveness*
    - *quality*
    - *topicality*
    - *reaction speed.*

On the human side, one can add security, group identification, and two sub-divisions of self-actualization: job richness and individual influence. How delegation, participation and aggregation differently affect these objectives is summarized in Fig. 5.1.

However, in work planning there is yet another dimension to consider. For work planning decisions can be grouped around six different issues: the setting of work standards, the determination of work goals, the administration of work, people and material resources, and the final, more general, issue of work improvement.

On each of these, the relative importance of the objectives listed above varies. It should therefore be expected that the desirable degree of delegation, participation and aggregation would also vary between the issues.

6

In the field of *compensation schemes*, there has in the last few years been a determined movement virtually everywhere for the stabilization of incomes. In most countries, this has manifested itself as a drive to eliminate piecework. In the UK especially, however, it has also been a drive to stabilize income irrespective of hours worked: to place blue-collar workers in a similar position to that common among white-collar.

The main motives for the movement have been that, on the human side, fluctuating incomes create insecurity, and that, on the economic side, pay packets that require complex calculation are more likely to lead to disputes and stoppages.

Unfortunately, what evidence there is is that changing to fixed wages from piecework systems leads to a drop in productivity. Best results seem to be achieved by systems which are relatively stable but incorporate a productivity-linked premium.

Closer analysis shows that the pay packet can be made up of payments for several different factors, each of which affects the objectives of the system differently. The ideal pay scheme in any situation will therefore be one that uses some combination of those factors in a way that optimizes the overall effect on the range of objectives involved.

Of the various factors, two are fixed. These are *pay for the job* (a fixed payment made weekly according to the job being done, and irrespective of output or hours worked), and *pay for the man* (a fixed payment made according to the worker's qualifications and circumstances).

Three are variable. These are *pay for time spent working* (paid according to a standard hourly rate, irrespective of output); *pay for contribution to output* (paid according to some measure of output) and *payment for company performance* (bonuses paid irregularly and according to the company's performance, but irrespective of the recipient's personal contribution to it).

The objectives that these different factors can be expected to contribute to are, on the economic side, flexibility, self-correction, stability, and performance; on the human side, they are security, group membership, individual influence, and self-development.

The way in which they act on the objectives, and therefore the key to a comprehensive and successful mix, is summarized in Fig. 6.2.

7

Over the years, the problem of meeting the economic objectives involved in housing the production system has been more or less conquered by the evolution of the 'multi-purpose box', in which virtually everything can be placed anywhere. A huge industrial hall at one level, with no partitions, a few pillars and preferably no windows, it has become the archetype of industrial building in modern times. It simultaneously achieves the key economic objectives of high productive capacity, flexibility and low cost.

Of itself, however, such a structure does nothing one way or the other for human objectives. Of those, the important ones involved are security and group membership. And they are affected much more by the situation inside the building, and the choice of location.

Internally, five basic components of the physical environment have to be controlled. These are: sound, light, heat, air and layout. The chief impact of sound, heat, and air pollution is on security, though providing the worker with the means to control his own working conditions, where possible, can also help with self-actualization needs. The same is true of individual light controls, though the chief impact of light is on the performance levels it makes possible. As far as layout is concerned, the human objective that has to be balanced against the economic need for efficiency is the need for group membership. In particular, the choice of layout needs to be compatible with the group structure envisaged in the production mode and work planning system.

Choice of site is normally limited by zoning restrictions. Where possible, however, considerable advantage can be gained for human objectives by establishing the plant as part of an integrated community, rather than in a single-purpose industrial development.

8

The last of the components that have to be considered in the sub-system phase is the *organizational environment*. By that we mean effectively the degree and way in which the individual worker is involved in the affairs of the company outside his own workplace.

In recent years, there has been a general move to increase the amount and kind of participation in this sense. This has been for three reasons:

— because it has been legislated for
— because some managers have believed it to be ethically right
— because it is held to increase industrial efficiency.

Only the third of these is something we can usually comment on.

In comparing organizational forms of participation, four factors need to be taken into account: their scope, their selectivity, the degree of participation, and its formality. We also need to bear in mind the distinction between participation and delegation: increasing participation does not necessarily carry with it the benefits attributable to decentralization.

For the *scope* of the system—the number and breadth of the issues covered by it—the chief limiting factors are likely to be the degree of specialist knowledge needed to make any particular decision, and the speed with which it needs to be made. The *selectivity* of the system—the number and kind of people it involves—can be based on any one of three principles. The first—participation on the basis of power—includes everyone who has a significant power to influence the outcome of the decision. The second—participation on the basis of involvement—includes all those who may be affected by the decision. The third—participation on the basis of qualification—includes all those who have the relevant and necessary skills and knowledge.

The *degree* of participation can be rated on a six-point scale. Workers may:

— make the decisions themselves
— share in the making of the decisions
— raise issues for decision
— be asked for opinions before the decision is taken
— be kept informed during the decision-making process
— be told about the decision after it has been taken.

Finally, the system can have a greater or lesser degree of *formality*, with varying effect and varying suitability in different circumstances.

An immense variety of organizational possibilities therefore exist. But there is an inherent tendency for the organizational system to parallel the structure of the management hierarchy. This is natural enough, and effective enough as long as the tendency to excessive bureaucracy is avoided.

From chapters 3 to 8, taken as a whole, it should emerge that no-one in the present state of the art can lay down simple and universal rules for the design of socio-technical production systems. What should also emerge, however, is that there is a general pattern to which it is possible to work.

And that is, first, to determine the particular set of economic and human objectives that each particular sub-system is intended to achieve. Secondly, it involves identifying all the components of the sub-system that can be independently designed. And thirdly, it means looking at the individual effects of all those components to ensure that all the desired objectives have been taken care of, and no undesirable side effects have been introduced.

9

Even when the production system has been designed, or redesigned, it still has to be introduced, and made to work. And that means, in present-day circumstances, introducing it in a way that commands the support of the workforce.

Once upon a time, in Generation 1 of change management, that was relatively unimportant. Changes could be implemented in an authoritarian way. Even if they were not accepted wholeheartedly, there was little anyone could do to oppose them. This was the generation of 'decide and order'.

Later, the importance of winning people over to accept the changes was realized. Generation 1 gave way to Generation 2: 'decide and persuade'.

Now, however, even Generation 2 is proving inadequate. The need is for yet a third generation of change management: one in which the guiding principle is 'decide together'.

10

Deciding together demands that careful attention be paid to the organization of project management, and to the introduction into project teams of relevant and involved members of the workforce, as well as combining local management with specialists, whether they be internal staff, or outside consultants.

It also means improving the general receptivity of the environment to change and its ability to participate in it. That means educational programmes. It also means developing participative methods of coping with routine affairs in order to encourage the habits and skills of consultation.

That done, the desirable outlines of project organization are relatively clear. Reporting to the initiating authority, there will need to be a group of involved personnel with a chairman with overall uathority and responsibility for the project. That group may also need a less fully involved group with a wider background for reference and advisory purposes.

There will also need to be a full-time project leader, who may or may not be a consultant, with, reporting to him, a series of separate consultative groups for each significant aspect of the project.

This short summary of the argument concludes our book. As an argument, it is a complex one. In that, however, it merely matches its subject matter. For the business of attempting to build production systems that will satisfy the full range of economic and human objectives is a complex one.

Sometimes one gets a different impression. It is easy to make sweeping generalizations about human behaviour, both inside and outside industry. It is easy to come up with 'obvious' or 'fundamental' principles. But it is much harder to make the generalizations stand up to critical examination.

Participation is not always a beneficial influence: some people do not want to get involved. Piecework is not always an incentive to increased performance: and stable wages are no panacea. Assembly lines are not necessarily dull or soulless: and they are efficient. And so on.

So, if anyone reading the book was looking for a simple 'how-to-do-it' guide, he has been disappointed. But not as disappointed as he would have been if we had tried to provide one. Because it would not have worked.

Designing production systems, in the current state of society, has to be a complex business. And if this book has done something to help the reader realize the full extent of its complexity, it will have fulfilled its purpose.

Appendix: Some who tried

Two into one will *go. Back to the assembly line. The engineer's contribution. The tangled web.*

Fokker Drechtsteden

Two into one will go

Fokker is one of the most famous names in aviation history. The company that bears it took on its present form in 1970 as a result of the merger between the Royal Netherlands Aircraft Factories Fokker N.V. with the German Vereinigte Flugtechnische Werke. It is the first true international merger in the aerospace industry.

Its subsidiary, Fokker Drechtsteden, comprises three plants. The production programme of this unit now covers a fairly wide range of activities, but the mainstay still is, and will be, the production of major airframe components, wings, and fuselage sections.

Two Drechtsteden plants, the former Aviolanda factory at Papendrecht and the De Schelde factory at Dordrecht, are located less than 10 miles apart. Both were founded as independent aircraft factories and were acquired by the Fokker Group in the 'fifties. Then, in the 'sixties, it became clear that combining them into one integrated organization offered an opportunity to introduce a new approach in industrial organization, an opportunity which was taken by the new managing director, Mr Aris.

In looking at the pyramidal structure of the Fokker organization, he noticed that the various levels could conceptually be grouped into four according to the kind of work each was responsible for. At the top were the *initiators*, below them, the *organizers*, below them, the *controllers*, and, holding up the base of the pyramid, the *producers*.

Admittedly, this is not an unusual organization model: and obviously it was not as clear-cut as the sharp division makes it seem. To Mr Aris, however, it was also obvious that everyone had the capacity, to some degree or another, to work in each of these roles.

Would it therefore not be possible to transform the large pyramid into many smaller ones, where more people would be involved in all four levels and consequently find more satisfaction in their work? Obviously, a change would be difficult: the present organizational set up had worked for many years and had performed quite well in many areas.

There was a further reason for reorganization, important even if the other two had not existed. Over the years, the systems of production planning and scheduling—as such systems frequently do—had grown cumbersome and slow. Mean lead time of quite simple details had increased to 10–14 weeks, and it was evident that the whole system was suffering from an excessive burden of red tape and paper work.

The reorganization programme was therefore initiated with three goals in mind: rationalizing the two factories into an integrated whole, simplifying the management control systems, and providing greater job satisfaction not only by diminishing the pyramidal distinction between roles, but also by opportunities for job rotation.

The entire programme was a complex one. For our purposes, therefore, we will focus our attention on one section of it only: the sheet metal details shop, which at Dordrecht had 80 employees, and at Papendrecht (where the new integrated shop was to be sited) 40.

To begin with, an interdisciplinary working group was set up to design the new shop, with the objectives of reducing lead time, improving job satisfaction, and simplifying the administration. Outside help was obtained from the Delft University of Technology, which provided a graduate consultant to carry out research work and provide various data and design alternatives for the working group's decisions.

The group was also assisted by the setting up of special consultation groups of factory personnel, to advise on general aspects of design, and, in some cases, even to provide the detailed layout of their own workplaces. During the course of the programme, student sociologists from the University of Rotterdam were also used to interview workers about their views and reactions.

The advantages of using group technology are well known, and this principle of good engineering practice was applied to the sheet metal details production. The number of operations on these parts is limited, usually five or less, and using fairly simple methods it became clear that it was possible to group all products in a limited number of product families. These product families were then allocated to so-called 'production streets' and estimates were made about the required capacity for each street.

As a result of this study, the machinery was regrouped into 10 production streets, each producing one main product family. Two of these streets manufacture in fact the same product family. The work was divided between two production groups, to keep the maximum number of people in the group below 15. The layout of the shop embodies a quite adequate material flow.

In principle, a regrouping such as this requires additional machinery. In this particular case, however, the company was able largely to ignore this problem: since two shops were put together in one, there was a surplus.

At present the new factory covers about 3700 sq. in. of floor space. The output of the details factory is now about 600 shop orders per week. The workforce totals about 110 people.

Administrative procedures were simplified considerably. A computer is used for workload planning. Every week an advice list per cost centre is provided which states the products that should be manufactured in the coming weeks. Feedback is limited to three events; the time that materials are taken from stock, the time when the parts enter the common surface treatment shop, and the time when the parts have passed the last quality control check and are sent to the finished parts store. Between these three events, the working groups administer their own production autonomously.

The direct results of these changes are considerable. Lead time for sheet metal parts was reduced to four weeks from 10. The strength of the scheduling department was reduced by 25 %. There is an overall productivity increase estimated at about 10 %, and the number of levels in the management hierarchy of the factory has been reduced from six to four.

In all, the operation took about a year of preparation and design work, and nine months of gradual implementation.

The programme has led to the following observations and conclusions on the part of the Fokker management:

- The building of a new organization is an evolutionary process: the new norms, procedures and agreements take time to become part of the instinctive mental equipment of the company population, which has to happen before the new organization becomes a living entity.
- It is important that consultation takes place with open attitudes. To this end, the company has regularly published a special journal giving news about progress and problems in the reorganization, and inviting criticism.
- Changes in management philosophy should not be introduced as responses to *present* problems. Doing so leads to defensive reactions. Change is only sensible and appreciated as a reaction to foreseen future events. Adopting this attitude promotes creativity, and by adhering to it the Fokker management very largely avoided arousing defensive attitudes, despite the often very lively discussions and the introduction of very significant changes.
- When subjects to be studied are largely covered by the same discipline, consultation groups should not consist of too many members. The group then becomes cumbersome and does not function. While this is a simple truth, it is nevertheless not easy to adhere to.
- Uncertainties about changes ('Where will I be working? and who with?') cause more uneasiness than the changes themselves, and diminish

interest in participation. This is essentially in line with Maslov's needs hierarchy—while security needs are unsatisfied, little interest in higher level satisfaction is likely to be manifested.

The final question concerns the actual experience of the participants on the shopfloor. More autonomy and responsibility, coupled with job enrichment should result in better job satisfaction and these results should be measurable. Progress in this area is steady and gaining momentum.

It was at once realized that an extensive training and instruction programme would be required, and such a programme was established for the management of the shop at the very beginning of the reorganization. This programme is now being widened to cover the working groups themselves and the first results are encouraging.

The opportunity for job enrichment by gradually learning to perform all work in one group is better received by younger people than by older; only 30 % of all workers have indeed significantly widened the scope of their work. Also it is necessary that the job and wage structure supports this job rotation, and this system is at present under revision.

Much more is to be done, but, with the technical and management control objectives already attained, Fokker is confident that the results in the field of job enrichment will also be good. This process is now in full swing and the next few years will witness a continuing delegation of authority to the working group.

Danish State Railways, Tåstrup

Back to the assembly line

It has to be admitted that 'back to the assembly line' is a slight misnomer on two counts. In the first place, this case is one concerning a maintenance operation, not manufacturing, so assembly is inaccurate. And, in the second place, 'back to' is perhaps misleading since the enterprise concerned had never had an assembly line (or 'maintenance line') before.

But the misnomer is justifiable in spite of that, as the case itself will explain.

Whatever may have been happening to railways in other parts of the world, the Danish State Railways (DSB) network that serves Copenhagen and its suburbs has been experiencing a boom in demand.

Whereas in 1960 it operated some 260 coaches, by 1974 the figure had risen to 500, and it is expected to reach some 800–1000 by 1980. By 1962, the management of DSB had already decided that this growth would demand the erection of a new workshop for the maintenance of rolling stock. In 1970, plans were approved, and in 1974 the plant was opened, at Tåstrup.

A railway coach is a fairly large item, and when it is out of service (especially without its wheels) it is relatively immobile. Traditionally, therefore, the maintenance worker has been brought to the coach. The production mode, in the terms of chapter 3, has been a stationary one. Unfortunately, rotating the workers among the carriages means that they have to use mobile equipment, which means, in turn, that they are unable to make use of the most modern and efficient equipment available.

In designing the Tåstrup plant, therefore, it was decided to move the coaches, not the men: to change to what we called in chapter 3 a *discontinuous production flow*. Separate workplaces were set up for the different maintenance tasks involved (painting, renovation of electrical equipment, wheels, brakes, motors, and so on), and the coaches moved from one to another in turn. Since, as we pointed out, railway coaches are large, keeping buffer stocks between workplaces is impractical. One work station cannot take in a new coach until the next station is ready to receive the one it has just finished. For maximum efficiency, therefore, all the coaches in the plant need to move at the same time, and the operations needed at each workplace have to be of approximately the same duration.

To all intents and purposes, therefore, we are back with the assembly line. Which justifies the misnomer with which we opened the case study.

The technical problems involved in balancing a line of this kind are considerable, especially since the amount of work needing to be done on each coach is unpredictable. But they were solved. Under the old system, the amount of working time theoretically necessary to service 500 trains a year was 130 000 hours. That would have been with perfect utilization, however. At the level of utilization actually achieved, the time required rose to 195 000 hours.

In contrast, introducing the new methods reduced the basic time required to 80 000 from 130 000 hours. Simultaneously, the utilization rate was also improved. With down time, other breaks, and idle time allowed for, the number of hours needed was reduced from 195 000 to 100 000.

Even coupled with the utilization rate, such figures remain theoretical, however. The amount of time needed in practice also depends on the performance ratio. In the old system, that ratio was about 0·72. The theoretical 195 000 hours therefore took in practice about 270 000 hours.

If the ratio remained the same, the new system's 100 000 hours would take 140 000 hours in practice. That would have been a saving of nearly 50%.

Nonetheless, considerable efforts were made to improve the performance ratio too, by paying attention to possibilities for improving the satisfaction of human objectives—despite the adoption of the 'assembly line'. In the first place, this was done by ensuring that the introduction of a discontinuous flow pattern was not accompanied by either method orientation or individual assignment. The traditional automatic assumptions were therefore avoided.

The entire process was too complicated to allow of a product oriented approach. A degree of craft orientation, however, was possible, within the limits set by the engineered workplaces, tools and quality specifications.

Group assignment was also used. At each work station, the workers were organized on a group basis, to the extent that they even elect their own leaders. On each task, they work together without any of the traditional demarcation barriers (the system was introduced on the basis of a special agreement with the relevant trades union).

The work analysis that had to be done to establish and balance the line was also put to further use. For the first time in DSB, it was used as the basis of a premium wage system, again based on agreement with the union. Each work group which over a three-week period averages 105% of its performance standard receives a premium bonus of 14% of the basic wage.

While method and utilization ratios can be precalculated, for performance one must wait and see. As the system has not yet been in operation long enough for a final verdict to be lodged, we must accept only indications of potential performance. Latest measurements are that the performance ratio is about 1·00, and there is every sign of further increases.

In terms of staff, that would mean a 63% reduction in the staffing requirement. What that means in terms of money is shown by the table (figures are in millions of Danish crowns):

	Old system	*New system*
Performance ratio	0·72	1·00
Cost of facilities	40	40
Cost of equipment	8·1	9·6
Project personnel costs	0·5	1·5
Consultancy costs	0	0·8
Total investment	48·6	52·9
Gross annual wage cost	6·8	2·8

The advantages of the 'return to the assembly line' are obvious and the economic benefits have been clearly coupled with the benefits of more interesting, cleaner and less tiresome work. While sweeping changes in organization, methods, wage scheme and output were introduced, they were all met by the workers in a responsible and committed way. But then there were no surprises for them or their organizations. From the start of the programme, their representatives were involved actively in the design of the new system, for which they were extensively trained.

Saab-Scania Södertälje

The engineer's contribution

The story of the development of work groups at the Saab-Scania's truck and

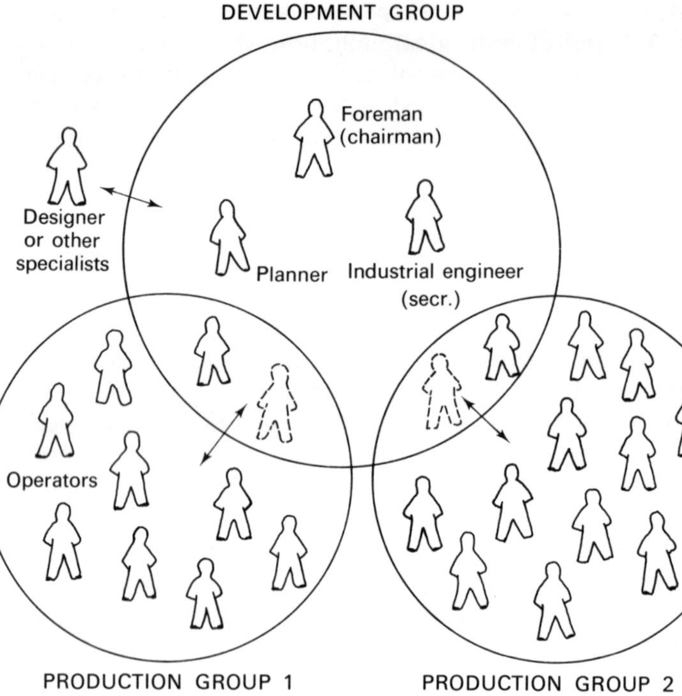

Fig. 12.1.

engine plant at Södertälje is a fairly well known one. Not always emphasized, however, is the critical role played by industrial engineering in making the development of the full potential of the system possible.

The Södertälje story started in 1969, when the management, faced with a long string of problems of recruitment, high personnel turnover, absenteeism, friction between managers and workers, and so on, decided to experiment with new ways of structuring work. Specifically, they set out to increase the possibilities open to each employee for influencing his own work situation, to make jobs more attractive and meaningful, to reduce susceptibility to disruption, and to increase production efficiency through greater flexibility and adaptability.

To begin with, they experimented. A 20-strong reference group, chaired by the technical director and including representatives of all the functional groups likely to be involved, was set up. They selected for the initial area of operation two departments, engine finishing and small-bore piping, employing about 40 men between them. In each of the two departments a work group was set up (including the supervisor, some fitters, and work study personnel), and asked to present suggestions and ideas for a modified organization. The fact that the experiment was taking place in an established workshop obviously limited the technical possibilities for change, but it was nevertheless felt that useful knowledge and experience could be gained.

The result of the experiment is well enough known to warrant only brief treatment. Instead of operating their traditional tasks as individuals, the workers were formed into production groups, each with a straightforward production goal. They were formed in such a way that all members would have real need of contact in their daily work. The groups were given greater responsibility for maintenance and quality control, and members were encouraged to train themselves for rotation between group tasks.

In addition, development groups were set up to provide greater opportunities for consultation on questions of routine production between supervisors, workers, industrial engineers, designers, and so on. On average, there was one development group for each two production groups. Each was chaired by the foreman, with a production technician as secretary, representatives of each production group, and whatever particular specialists might be required for the discussion of any specific problem.

This system is illustrated in Fig. 12.1.

In general, the innovations were seen to be highly successful at meeting the problems that had given birth to the project. The system was extended until it covered 1500 workers instead of the original 40. However, as we have pointed out, there were implicit limitations on what could be done within the confines of an existing factory. The project therefore moved to a more interesting stage when it became necessary to set up an entirely new production system at Södertälje for the production of the new Saab 99 engine.

Facilities for operating production groups were to be built into the system from the beginning. On the other hand, 110 000 engines were going to have to be produced economically each year. Some sophisticated industrial engineering thinking was therefore called for.

Those responsible for producing that thinking were provided with both human and economic objectives. From the economic point of view, they were required to meet the target production volume within specific investment limits, and at a maximum production cost per engine. And they were to fit the system into a pre-set floor area.

On the human side, they were asked to:

— reduce the extent to which the individual is tied to his job
— increase his opportunity to develop with his job
— provide opportunities for him to influence the distribution of work and his own work conditions.

The initially favoured solution revolved around keeping the assembly line,

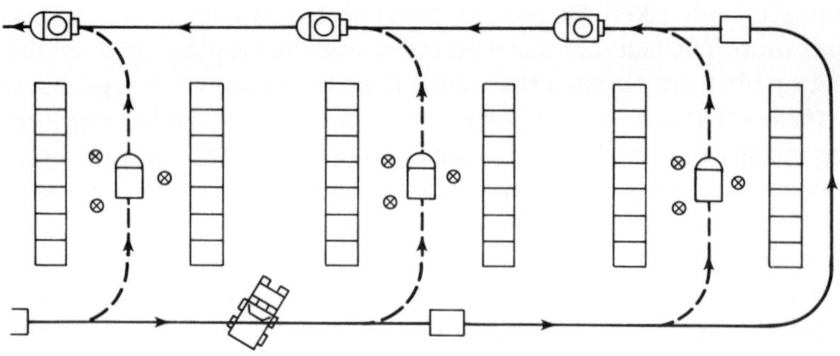

Fig. 12.2.

but providing buffer stocks at several points to enable the workers to determine their own work pace, and at the same time to allow work performed between the buffers to be arranged on a group basis.

This, however, was discarded on the basis that it did not allow sufficient variety of tasks in the production groups. The only thing that would do that, it was decided, was to ask each production group to produce a complete engine. The various groups would then work in parallel, each being provided with the unassembled components, and producing a finished engine out of them, with freedom to determine their own work patterns within that task. Instead of the 1·8 minute cycle time that each worker would have had to meet on a conventional assembly line, each group would be given a standard time of 30 minutes to produce the whole engine.

The problem remained how to deliver the components to the groups and how to take away the engines when they were assembled. The first solution proposed was that of Fig. 12.2. A U-shaped conveyor loop could be built around the entire line of production groups. Group members could take components off one leg of the loop, assemble them, and put the assembled engines on to the other leg.

The trouble with that was that the trucks bringing tools and materials to the racks would have to cross the conveyor belt. It would have been technically feasible to install a partly overhead conveyor which would cross the transport ways at sufficient height to allow the trucks to pass underneath, but, on investigation, it proved impractically expensive.

The final solution was one of those that looks obvious—afterwards. In fact, it took some while to produce it.

Figure 12.3 shows it schematically. Instead of passing the U-shaped loop totally around the line of production bays, the loop was kept completely to one side of them. Each group was then furnished with a U-shaped guide track in the floor, enabling them to take components as required off the side of the

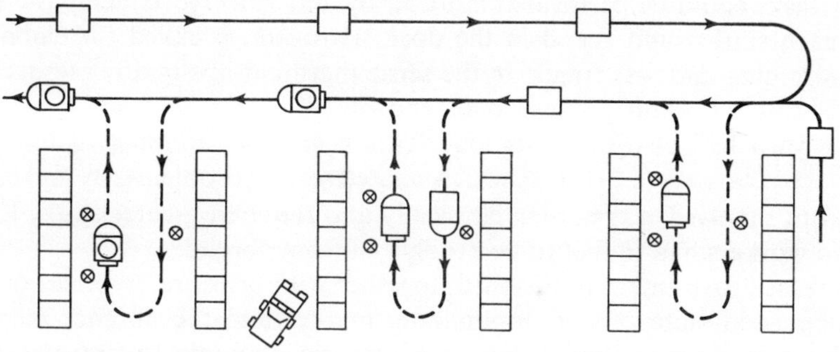

Fig. 12.3.

loop nearest them, and replace the finished engines on the same leg. Meanwhile trucks had completely unhindered passage from the other side.

This system was the one implemented, and is now functioning. In all, there are seven production groups (one of them is intended as a training group), and most of the original targets have been met, though the factory is still in the commissioning stage.

Personnel turnover was reduced to 20% in the chassis shop in 1972: in the same year, assembly costs were 5% under budget.

Certainly, Saab-Scania is now in a position to reap the full benefits of the production group system in a way that it could never have been while it was still limited to an existing factory. And in a way that it could never have achieved without a fair amount of imagination and flexibility on the part of the industrial engineers involved.

Kockums Mekaniska Verkstads Malmö

A tangled web

The history of Kockums shipyards in Malmö over the last few years provides no clear-cut lessons for deciding what changes in organization are likely to produce what results. However, in that it is typical. The whole story presents an excellent example of the difficulties facing anyone trying to derive universal principles from the study of real-life cases. The many factors involved, some under the control of management, some under worker control, and some under no control at all except that of the market and the outside world, create a tangled web of relationships that it is possible to unravel in many ways, none of which can lay any indisputable claim to being the right one.

Through the 1960s, many changes had taken place in the shipyards. Specialists originally in the building of supertankers, the firm had developed over the decade an assembly line method of construction. This meant that many tasks could be performed more easily and quickly, reducing the time that each ship would spend in the dock. However, it called for elaborate programming and was fragile in the sense that hold-ups in any one section could bring the whole process to a standstill.

The yard had also made its first venture into the building of liquefied natural gas carriers (LNGCs), which created several problems. By and large, the work involved was harder physically and required greater skills. Extra skilled workers had in fact to be brought in from abroad.

Increased investment in the yard, together with pressure on prices due to the depressed state of the shipbuilding market and the absence of state subsidies in Sweden, had led to considerable attempts to improve productivity through traditional means. Extensive use was made of industrial engineering and, partly as a result, piecework rates were changed and tightened. In particular, in some major departments, the switch was made to basing piecework rates on the time tasks *ought to* take rather than, as before, on the average times they actually took.

In 1967, time clocks were introduced in all departments, and attempts were made to cut down on coffee breaks and stop workers leaving early.

Ever since the late 1940s Kockums had been importing foreign labour. Increasing tightness in the labour market in the area also drove them to greater use of so-called 'grey' labour, hired out to them by independent entrepreneurs. By 1969, 1000 of the yard's 3500 workers came from this source.

In that year, Kockums management, faced with increasing labour turnover and discontent, especially among the key welders and platers (in 1968, 688 platers and welders left the company, which only managed to recruit 650 others in their place), took an unprecedented step. They asked the local Metal-Workers' Union to carry out a study of their workforce to establish the reasons for the dissatisfaction. The union, short of time and resources, passed the request on to the Swedish Labour Organization (equivalent to the TUC or the AFL-CIO) who accepted it. Believing that the study would be of use to their own organization, the SLO even agreed to share the costs with Kockums. A probably unparalleled study of worker attitudes then took place, intended to produce a report expressing worker opinions directly and fully and diagnose the reasons for them.

In the eventual report, the highest ranking of those reasons was dissatisfaction with economic rewards, though this could be divided under several heads:

— Under a strict piecework system, earnings varied directly with the productivity of the worker, but that productivity was not under his control. Poor work planning and scheduling, inadequate supplies of tools and materials, all had a direct effect on the worker's earnings, though there was nothing he could do about them.

— The sensitivity of the assembly line system to disruption had led to a diminishing of demarcation between tasks and a flexible use of labour. This raised the efficiency of the production system as a whole, but lowered the worker's own efficiency. Each change of job could lose him money.
— That same flexibility also meant that workers were frequently shifted from one work group to another. Since most piece-rates were on a group basis, this also meant that the worker's earnings were dependent on the efficiency of colleagues who might change from day to day, with many of whom, in view of the high proportion of foreign labour, it was difficult to communicate.
— The new piecework systems that had been introduced were highly complex and difficult to understand, and led to disputes.
— The situation favoured some work groups—those with high solidarity and a good knowledge of the system. Thanks to their skill at 'systematic soldiering' they were able to exploit the system. But this led to jealousy on the part of those less well placed.
— Jealousy was also created by the fact that the 'grey' labour was employed on a different basis, being paid by the entrepreneurs who hired them to the firm. Not only did they in some cases get higher allowances, but, in general, there was a much greater fixed component in their pay. (The foreign carpenters were, in fact, paid fixed wages.)
— Finally, the tightening of rates that had occurred in some areas predictably raised resentments.

Social conditions were also an important source of discontent, though the picture here was not as clear-cut. Relationships between workers and lower levels of management were actually found to have improved. This was due to the fact that foremen and supervisors were no longer responsible for rate-setting. Antagonism to the 'experts' bound the lower levels together, and, on the whole, foremen were prepared to help workers gain better rates in order to keep them at Kockums. In some work groups, solidarity was high, though the increasingly flexible use of labour had done much to break this up. In general, most workers felt that increasing instability of the work group heightened their insecurity.

The growing intake of foreign labour and 'grey' labour also weakened labour solidarity, and, in the case of the foreign labour, most of it from Finland and Yugoslavia, increased difficulties in social communication.

Finally, the changes that had occurred, and the way in which they had been introduced, together with the stricter supervision resulting from the attempts to increase productivity, had generated a deep distrust of senior management.

Health and welfare problems provided only a minor source of discontent, though complaints about the effect on health of shiftwork were fairly common. There was concern about the way in which tighter piecework rates increased the pressure to take risks on the job—for instance, by erecting scaffolding too hastily. Some complaints were made about poor ventilation and occasional overcrowding, and some workers claimed that the irregular work flow and the pressures of piecework produced nervous stress.

As far as work satisfaction was concerned, the report was, by contrast, positive. The introduction of the 'assembly line' had brought none of the problems usually associated with it. Tasks remained fairly large in scope and meaningful to the worker. Much of the work still retained satisfying craft characteristics.

Moreover, product identification was high. It was easy to follow the progress of the building of each ship, and the workers could easily recognize their own contribution to it. The publicity and ceremonies surrounding the launching of each vessel also produced a favourable effect.

The heaviness of the work, the occasional danger of it, and the exposure to all weathers involved in working in the dock worried a few people, but, on the whole, the workers considered themselves fortunate from the job satisfaction point of view. However, as the report pointed out, workers generally tended to pass very quickly over this subject to get on to what they considered the important matters.

There were therefore many reasons for discontent, and, not unnaturally, the management adopted several measures to counteract them. Since the study, the changes that have occurred include:

- Straight piece rates have been abandoned, and replaced by a premium system with a large fixed component. 'Grey' labour is now paid directly by Kockums in the same way as the rest of the labour force.
- Various services, notably the supply of materials and tools, have been decentralized, with the result that the worker is less dependent on the working of the rest of the system for his own (or his work group's) productivity, to which the premium is linked.
- The statutory works council now devotes more time to personnel matters, departmental councils have been set up and regular informal discussion groups introduced. All of these give the individual worker more chance to make his own views and feelings felt.
- A separate structure of groups has also been created to handle the submission of suggestions for improvements.
- The company's internal magazine has been enlarged, and a supple-

mentary fortnightly information sheet introduced, jointly produced by management and union. (Foreign workers are provided with foreign language editions.)
- The union has increased the number of representatives engaged full-time on wage and personnel matters.
- There are now two full-time doctors instead of one part-time one and medical services have been extended.
- The personnel department has been extended and includes a Yugoslav and a Finnish-speaking social worker.
- Attention to safety has been increased, many new training facilities have been introduced, two new hostels for unmarried workers are being built, and much more attention is being paid to recruitment procedures in the attempt to keep labour turnover down.

On the face of it, the changes have worked. Labour turnover and absenteeism have decreased sharply, productivity has increased and production speeded up. The need for 'grey' labour has been considerably reduced as a result, and the firm's profit situation is greatly improved.

About the only original problem still remaining is the difficulty connected with the assimilation of foreign workers, but this is something that much more powerful institutions than Kockums have failed to solve. Where governments fail, shipyards cannot be blamed for having problems.

Unfortunately, when it comes to pinning the credit for the improvements to particular changes, the task is virtually impossible. It would be possible to put the whole of it down to the elimination of piecework, and discount the other factors as trimmings. Equally, it would be possible to give the credit to the increased attention to participation, or the various 'human relations' improvements that have taken place. It would be logical enough to give them all credit. But no one explanation can claim to be exclusively demonstrated. Take a shotgun to shoot a bird, and it is very difficult to pick out the pellet that succeeded.

This is especially true when one tries to take into account some of the other things that happened, and that would have happened anyway. The shipbuilding market has gone through a boom period. The pressures for maximum productivity have therefore dissipated, and the whole industry (or most of it) has been benefiting from increased profitability. The yard has now grown accustomed to the production of LNGCs as well as tankers, and the problems associated with their introduction have disappeared. The general labour shortage in the whole Malmö area has eased, allowing reduction in the amount of 'grey' labour employed as well as lessening the inducements on workers to leave.

There is no way of adjudicating. As we pointed out, the web of relationships is too complicated for one answer to be acceptable. All one can do is look at all the problems there were, look at all the changes that were introduced, and ponder their significance against the background of one's own experience.

Index

AFL-CIO, 172
ABC-analysis, 68
Absenteeism, 6, 7, 12, 17, 34, 83, 168, 175
Accountants, 145
Achievement, 12, 25, 26, 94
Advisory committee, 150
Affluence, 137
Agent of change, 141
Aggregation, 55, 58, 60, 61, 63, 64, 72, 73, 156, 157
Agreement on rationalization, Swedish, 5
Air-conditioning, 107
Aircraft industry, 162
Akerman, Johan, 128
Alcan Aluminum, 6
Alienation, 81, 153
Amalgamated Union of Engineering Workers, 5
American Institute of Industrial Engineers, 85
Architects, 103, 104, 107
Aristotle, 59
Army, 89, 90
Assembly costs, 171
Assembly lines, 3, 12, 28, 29, 32, 38, 39, 43, 50, 51, 53, 54, 155, 156, 161, 165, 166, 169, 170, 171, 173, 174
Australia, 114
Austria, 130
Autocracy, 64, 73
Automation, 50, 58, 75, 110
Autonomous groups, 29, 39, 55
Autonomy, 45, 63, 138, 164, 165
Aviolanda, 162

Batch production, 38, 39, 52
Behavioural science, 18, 23, 24, 35, 80, 138, 154
Belgium, 130
Bismark, 2
Blue-collar blues, 10
Bonuses, 86, 87, 95, 96, 97
Breweries, 111
Britain, 2, 3, 6, 10, 13, 18, 24, 80, 82, 88, 89, 90, 112, 157
British Leyland, 82
Budapest, 18
Budgets, 128, 129
Buffer stocks, 166, 170
Building materials, 74
Bureaucracy, 121, 123, 160
Byron, 6

California, 80
Camus, Albert, 5
Capital costs, 92
Capital resources, 30, 31, 32, 155
Centralization, 58, 71, 72, 115, 125
Chaplin, Charles, 5
Charisma, 24
Chesterton, G. K., 6
China, 79
Civil Service, 89, 90
Class solidarity, 12
Colour, 106, 110
Colour television, 54
Community service, 12
Compensation, 33, 155, 157
Comprehensiveness, 61, 157
Computers, 70, 164
Conference board, 80
Constraints—economic, 42
Consultants, 140, 151, 160
Consultation, 169
Control, 61, 63, 91, 127, 163
Control power, 157
Controllability, 42, 44, 46, 48, 156
Conway, James, 5
Co-ownership, 96
Cost centres, 164
Cotton coats, 52
Craft-orientation, 45, 46, 47, 54, 167
Culbert, Samuel A., 12
Cummins, Julian, 14
Curtis-Wright, 57
Customer service, 76
Cycle time, 21, 22, 29, 39, 50, 51, 170
Czechoslovakia, 18

Danish State Railways, 165, 167
Data base, 142
De Gaulle, Charles, 6
De Schelde, 162
Decentralization, 58, 59, 61, 64, 156, 159

Decision, 139
Decision-making, 72, 112, 113, 115, 116, 117, 119, 126, 127, 150, 159
Defective hearing, 105
Degree of involvement, 115, 117
Degree of participation, 159
Delegation, 58, 59, 60, 61, 62, 64, 72, 74, 75, 77, 115, 123, 156, 157, 159
Delft, 163
Democracy, 59, 117
Denmark, 4, 10, 22, 50, 53, 99, 134, 149
Department of Employment, UK, 5
Design procedure, 144
Development groups, 168
Diagnostic power, 157
Diplomacy, 116
Direct labour costs, 100
Disruption, 168
Dividends, 96
Division of labour, 1, 6, 7, 8, 153
Divisions, 127
Dordrecht, 162, 163
Downtime, 166
Drucker, Peter, 23
Dual hierarchy, 149, 160
Dust pollution, 109, 110

EDP, 70
Eastern Europe, 79
Economics, 112
Education, 125, 127
Education gap, 9, 134, 137
EEC, 112
Effectiveness, 119
Efficiency, 119, 159, 173
Efficiency experts, 137, 138
Efficiency imperative, 117
Elden, James M., 12
Energy shortage, 15
Engineering, 127
Engineering industry, 99
Engineers, 145
Esteem, 25
Ethical imperative, 117
Ethics, 112, 114, 159
Etiquette, 119, 120
Europe, 96
Exhaustion, 105
Experience increments, 98

Facilities administration, 65
Factories Acts, 2
Fallacies, 28
Fayol, 19
Fielden, John, 2
Finland, 4, 130, 173
Fixed pay components, 87, 97
Fixed wages, 65, 79, 83, 85, 94, 157, 173
Flash removal, 109
Flexibility, 42, 44, 46, 48, 55, 87, 89, 90, 92, 93, 94, 95, 98, 106, 156, 158, 168, 173
Flow-through, 55
Fokker, 44, 164
Ford, 39
Foreign labour, 172, 173, 175
Foremen, 29, 94, 173
Formality, 159
Foundries, 109
France, 5, 96, 130
French Revolution, 8
Futurists, 5

Gantt, 19
General Dynamics, 57
General Election, UK, 1974, 18
General Motors, 7
Generation 1, 134, 135, 160
Generation 2, 134, 135, 138, 139, 160
Generation 3, 138, 139, 148, 160
Germany, 51, 131, 146
Germany, West, 5, 6, 14
Gilbreth, 3, 19, 137, 154
Glass, 127
Goddard, Paulette, 5
Graduates, 90
Gränges, 127, 128, 129
Grinding, 107, 109, 110
Group identification, 43, 49, 156, 157
Group incentives, 55, 93, 98, 100
Group inertia, 121
Group membership, 87, 89, 90, 158
Group Technology, 44, 46, 68, 163
Groups, 29, 32, 47, 48, 50, 51, 53, 55, 60, 74, 75, 101, 110, 111, 121, 156, 164, 174

Haraszti, Miklos, 18
Harvard Business Review, 12
Hawthorne, 18, 154
Hawthorne effect, 85
Health, 104, 174
Heat, 104
Herzberg, Frederick, 7, 26, 35, 111, 136, 138, 154
Hierarchy, 114, 115, 117, 119, 120, 121, 123, 141, 149, 151, 160, 164
Hierarchy (participative), 121, 122
Hierarchy of needs, 25, 165
Hill, Andrew, 6
Hire-purchase, 96, 97
Holland, 73, 131
Hollywood, 137

Hong-Kong, 15
Human relations, 24, 25, 26, 134, 136, 138, 175
Human resources, 30, 31, 32, 155
Hygiene, 111, 136

ICI, 80
Ideal-centred, 143, 144
Identification, 35, 42, 47, 48, 49, 61, 62, 63, 96, 106, 119, 120, 126, 132, 156, 174
Ideology, 17
Imbermann, A. A., 12
Impact of pay components, 93
Implementation, 132, 140, 164
Incentive schemes, 79, 81, 82, 85, 89, 92, 94, 95, 137
Indifference, 126
Indirect labour costs, 100
Individual assignment, 166
Individual influence, 62, 63, 87, 89, 92, 157, 158
Induction, 148
Industrial engineering, 3, 19, 22, 136, 137, 141, 146, 153, 168, 169, 172
Industrial Revolution, 1, 3, 6, 8, 16, 153
Industrial sabotage, 7
Industrial Society, 24
Inertia, 121
Information-processing, 71
Innovation, 139
Input planning, 58
Insecurity, 11, 35, 81, 135, 157
International Publishing Corp., 7
Investments, 128
Involvement, 116, 117, 138, 140, 160, 164
Irritation, 105

Japan, 15, 80, 90, 96
Job classification, 97
Job design, 37
Job economy, 134
Job enlargement, 55
Job enrichment, 26, 32, 55, 119, 165
Job richness, 62, 63, 135, 157
Job satisfaction, 165, 174
Job shops, 44, 46
Job simplicity, 42, 46, 48, 156
Job simplification, 135
John Lewis Partnership, 113

Kay, Dr J. P., 2
Key objectives, 93
Keynes, Lord, 15
Kockums Shipyard, 34, 171
Kodak, 107

Labour turnover, 6, 7, 34, 83, 124, 168, 171, 175
Laissez-faire, 2
Lapland, 127
Law, 114
Lawless, John, 13
Lawrence, D. H., 6
Lay-off pay, 89
Layout, 107, 108, 158, 163
Lead time, 163, 164
Legislation, 112, 113, 114, 115, 124, 130, 159
Light, 105, 106, 110, 158
Light contrast, 105
Light intensity, 105
Lindholm, Rolf, 79, 119, 120
Line sidings, 53
Location of plants, 159
Lordstown, 7
Lorries, 51
Lupton, Professor, 81, 82

MOST, 66, 68
MTM, 22, 66, 68
Machine utilisation, 72
Machine-pacing, 50, 51
Maintenance, 165, 169
Malmo, 171, 175
Management of change, 160
Management of projects, 141, 146
Management science, 138
Management style, 24
Manchester, 81
Mandle, Ian, 13
Market coupling, 40
Marxism, 2, 3, 113
Maslow, Abraham, 25, 35, 154, 165
Mass production, 38, 39
Mater et Magistra, 16
Material costs, 81
Materials administration, 65, 75, 157
Matrix organization, 149
Maynard, 118
Maynard, Dr, 19, 154
Mayo, 18
McLelland, 25
McGregor, Douglas, 26
Measured day work, 85
Metals, 127
Method orientation, 45, 46, 47, 54, 155
Method ratio, 20, 21, 35, 42, 87, 167
Methods engineering, 79
Military maintenance, 22
Mill, John Stuart, 2
Mismatch of perception, 119
Missile guidance, 22

Modern times, 38
Modular principle, 54

New employees, 148
New Lanark, 113
Noise, 104, 106, 110, 111
Norms, 164
Now-centred, 143, 144

Objectives, 36, 41, 48, 60, 66, 89, 125, 141, 142, 145, 146, 156, 157, 158
comparing, 147
economic, 41, 43, 60, 86, 89, 104, 154, 155, 156, 157, 158
human, 32, 34, 35, 42, 61, 78, 86, 89, 104, 154, 155, 156, 157, 158, 166
physical, 33
productivity, 32
Oil industry, 43
Organization theory, 122
Organizational environment, 33, 155, 159
Output planning, 58
Overtime, 89, 91
Owen, Robert, 113
Ownership, 113
Oxelosund, 128, 129

Package solutions, 39, 58, 155
Papendrecht, 162, 163
Paper-making, 97
Pareto's law, 68
Participation, 9, 10, 58, 59, 60, 61, 64, 66, 72, 73, 74, 75, 77, 104, 110, 111, 112, 113, 114, 115, 116, 118, 119, 120, 121, 124, 125, 126, 127, 133, 139, 140, 142, 145, 148, 149, 156, 157, 159, 161, 169, 175
Paternalism, 24
Pay components, 86
Pay for output, 87, 92
Pay for the job, 87, 88, 89, 97, 158
Pay for the man, 87, 89, 90, 91, 98, 158
Pay for time, 87, 88, 89, 91
Pay systems, 78, 157
Payment by results, 80, 81, 86, 87
Pennsylvania, 80
People and computers, 71
Performance, 87, 129, 158
Performance ratio, 20, 21, 23, 35, 42, 87, 166, 167
Persia, 16
Personnel, 128
Personnel administration, 65, 75
Persuasion, 127, 139
Physical environment, 33, 103, 104, 105, 155
Physiologists, 104
Piecework, 80, 83, 85, 88, 99, 100, 135, 137, 157, 161, 172, 173, 174, 175
Pilkingtons, 24
Pink refrigerators, 71
Planning, 57, 112, 128, 163
Planning and computers, 70
Plant location, 108
Poland, 18
Political imperative, 117
Pollution, 104, 105, 107, 158
Pope John XXIII, 16
Pork processing, 21, 22
Power, 116
Premium payments, 83, 85, 95, 97, 98, 100, 157, 174
Process production, 38, 58
Product identification, 43, 49, 156, 174
Product-orientation, 46, 47
Production flow, 40, 43, 46, 156, 166
Production groups, 169
Production mode, 32, 33, 38, 48, 58, 59, 68, 106, 155, 156, 158, 166
Production schedules, 65
Production types, 38
Productivity, 4, 10, 14, 15, 34, 81, 83, 85, 91, 94, 99, 100, 113, 135, 153, 154, 157, 168, 172, 174, 175
Productivity formula, 20, 21
Profit-centres, 127
Project groups, 74
Project management, 141, 146, 152, 160
Project manager, 149, 150, 151
Psychologists, 104, 145

Qualification, 116, 117
Quality, 61, 63, 83, 157
Quality control, 55, 169
Quality costs, 81
Quality of life, 14

Railways, 165
Rank, 57
Rationalization, 136, 163
Raw materials shortage, 15
Reaction speed, 61, 157
Realpolitik, 124
Receptivity, 142, 148, 160
Recognition, 26
Recruitment, 168, 175
Refa, 144, 145
Reference group, 150, 168
Representation, 117, 123, 128, 151, 175
Representation, 113
Resistance to change, 132, 133, 134
Resource phase, 30, 31, 32, 155

Responsibility, 165
Restrooms, 111
Reverberation time, 106
Rewarding input, 91, 101
Rigidity, 119, 120
Roethlisberger, 18
Rotterdam, 163
Rousseau, 6
Rubber boots, 50
Ruehl, Professor G., 14, 28

SMAC, 66, 68
Saab 99 engine, 169
Saab-Scania, 34, 44, 167, 171
Safety, 128, 129, 175
Salesmen's commissions, 93, 94
Scale of participation, 118
Scandinavia, 41, 89
Scania Vabis, 22
Scanlon Plan, 96
Schedules, 65
Scheduling, 163, 164, 172
Schuller Plan, 96
Schumpeter, 1
Schwab, 19
Scientific management, 18, 19, 21, 23, 24, 25, 27, 154
Scope of participation, 115, 159
Security, 87, 91, 92, 95, 132, 133, 137, 157, 158
Seeds of conflict, 13
Selection of projects, 141, 142
Selectivity, 115, 116, 159
Self-actualization, 25, 35, 43, 45, 46, 47, 87, 95, 106, 118, 132, 156, 157
Self-checking, 55
Self-correction, 87, 88, 91, 94, 158
Self-development, 26, 87, 158
Seniority, 80
Service industry, 76
Shareholders, 116
Shipyards, 30, 34, 43, 127, 171
Sickness, 129
Simplicity, 66, 68
Skill increments, 98
Smith, Adam, 1, 2, 3, 5, 8, 16
Social conscience, 114
Social contract, 108
Sociologists, 163
Sodertalje, 167, 168
Sound, 104, 105, 158
South Korea, 16
Spain, 131
St-Simon, 2
Stability, 87, 92, 93, 95, 132, 158
Standard methods, 64
Standard quality, 65
Standard times, 64, 65
Status, 25, 94, 119
Steelworks, 127
Steering committee, 149, 150, 151, 152, 161
Stegemorten, 19
Stock control, 31, 32
Stock-holding, 40, 44, 50, 52, 53, 164, 166, 170
Stock-options, 96
Strategic plans, 141, 142
Sub-system phase, 29, 35, 155
Sunlight, 106
Supertankers, 171
Survey phase, 29, 30, 155
Sweden, 4, 6, 9, 10, 11, 22, 34, 52, 74, 79, 82, 85, 94, 97, 98, 100, 110, 111, 122, 127, 129, 172
Swedish employers' federation, 5, 79, 82, 83, 119, 120
Swedish labour organization, 5, 172
System economy, 157
System phase, 29, 32, 33, 155
System power, 157
System resources, 30, 155
Systematic soldiering, 81, 82, 173

TUC, 172
Taiwan, 11
Targets, 129
Tastrup, 165, 166
Taylor, F. W., 3, 4, 18, 154
Teaching, 89
Teamwork, 148
Technological competence, 142
Teething troubles, 124, 132
Temperature, 107, 111, 158
Texas instruments, 7
The Times, 14
Theory X, 26, 138
Theory Y, 26, 138
Third World, 15
Timber industry, 100
Time study, 135
Topicality, 61, 157
Trucks, 167

USA, 2, 4, 5, 6, 10, 12, 23, 41, 80, 85, 96, 112
USSR, 79
Uncertainty, 164
Unemployment, 6, 11, 134
Union attitudes, 80
Union bargaining, 81
Union demands, 99
Union organization, 125

Unions, 3, 4, 116, 117, 167, 172, 175
Unit labour costs, 91, 92
Unit production, 38
United States Congress, 4
US War of Independence, 8
Utilization ratio, 20, 21, 35, 42, 87, 166, 167

Value-added, 96
Variable pay components, 87, 91, 98
Ventilation, 105, 107, 110, 174
Vilmar, Fritz, 17
Volvo, 22, 39
Vote-counting, 115

Wages, purchasing power of, 4
Wall switches, 52
War:
  First World, 3, 82
  Second World, 3, 6, 16, 96, 134
Welding, 107
Welfare, 174
Western Europe, 79
Wild, Professor Ray, 6
Wildcat strikes, 7, 79
Wilson, Dr N. A. B., 5, 7
Wimpisinger, W. M., 10
Woodward, Joan, 38
Work administration, 65, 75, 157
Work flow, 174
Work goals, 65, 75, 157
Work groups, 96, 104, 108, 118, 119, 120, 128, 129, 151, 163, 164, 165, 167, 168, 169, 170, 171, 173
Work improvement, 65, 76, 157
Work in America, 7
Work in progress, 31, 32, 55, 72
Work measurement, 79
Work orientation, 40, 45, 48, 156
Work places, 166, 167
Work planning, 33, 44, 57, 115, 155, 156, 158, 163, 164, 172
Work scheduling, 31
Work standards, 64, 66, 75, 157
Work stations, 21, 22
Work-ins, 9
Worker assignment, 40, 47, 156
Works councils, 174

Xerox, 57

Yugoslavia, 121, 173

Zoning, 109, 158